QUALITY MANAGEMENT
ESSENTIAL PLANNING FOR BREWERIES

By Mary Pellettieri

Brewers Publications
A Division of the Brewers Association
PO Box 1679, Boulder, Colorado 80306-1679
www.BrewersAssociation.org
www.BrewersPublications.com

© Copyright 2015 by Brewers Association

All rights reserved. No portion of this book may be reproduced in any form without written permission of the publisher. Neither the authors, editors, nor the publisher assume any responsibility for the use or misuse of information contained in this book.

Printed in the United States of America.

10 9 8 7 6 5 4 3 2 1

ISBN-13: 978-1-938469-15-2
ISBN-10: 1-938469-15-1

Library of Congress Cataloging-in-Publication Data

Pellettieri, Mary, 1968-
 Quality management : essential planning for breweries / by Mary Pellettieri.
    pages cm
 Includes bibliographical references and index.
 Summary: "This book decodes how to create and manage a system for ensuring quality in all facets of the brewery environment. Written for breweries of all types and sizes--new and established--this book will guide the development of a program that grows and changes with your brewery"-- Provided by publisher.
 ISBN 978-1-938469-15-2 -- ISBN 1-938469-15-1
 1. Breweries--Management. 2. Breweries--Quality control. I. Title.

TP570.P35 2015
663'.3--dc23
                    2015022558

Publisher: Kristi Switzer
Technical Editor: Dana Sedin
Copyediting: Christina Echols
Indexing: Doug Easton
Production and Design Management: Stephanie Johnson Martin
Cover Design: Kerry Fannon
Interior Design: Kerry Fannon, Justin Petersen
Cover Photography: John Johnston, Luke Trautwein, Kerry Fannon

This book is dedicated to my mother,
Georgeanne Pellettieri.

TABLE OF CONTENTS

Acknowledgments ix
Foreword by Ken Grossman xi
Preface xv
Introduction xvii

1. Defining Quality in a Brewery 1
    The Changing Definition of Quality Management 3
    The Gurus of Quality 3
  What Is Quality? 5
    Defining Quality as "Fitness for Use" 6
    Defining Beer Quality as Esoteric 7
  How to Achieve Quality – The Quality System 8
  Who – Quality Function in Breweries 9
    The Brewing Chemist 11
    The Chemist and the Quality Team 12
    The Change to Quality Is Everyone's Responsibility 12

2. Quality Management and Governance 15
    Connecting the Dots Between Management and Governance 15
  Quality Governance – Stating What Is Expected 16
    The Quality Manual – Setting Policy, Specifications, and Goals 17
    The Quality Manual – Roles and Responsibilities 19
  Quality Management – Implementation of Policy 21
    Quality Control and Quality Assurance 22
    Training 22
    Corrective Action 23
    Continuous Improvement 24

3. Components of a Quality Program 27
It's All About the People 27
The Front Line – Brewery Staff 28
Quality Skills for Brewery Staff 30
The Second Line – Middle or Upper Management 30
The Quality Staff 32
Assessing Skills and the Organizational Design 32
Conducting the Right Tests and Conducting Them Well 32
Conducting a Risk Assessment 33
Conducting a Risk Assessment with FMEA 36
Setting Specifications 37
Setting Limits 39
Setting the Frequency of the Check 41
Ease of Access to Data 43
Assuring the Test Results 44

4. Supporting Functions to the Quality Program 47
Human Resources 47
Asset Care or Maintenance of Equipment 51
Sanitation and Good Manufacturing Practices 54
GMP Culture and Implementation 55
Sanitation Planning 56
Validation 57
Record Keeping 57
Complaint Records 57

5. Strategic Components in the Quality Program 59
New Product Design and Implementation 59
Managing Innovation and Resources 61
New Product Introduction 62
Implementing a Structured Problem-Solving Program 64
Managing Problem Solving 66

6. The Best Tests for a Brewery 69
The Brewery Laboratory 70
Microbiological Tests 70
Microscopy 71
Cell Staining 72
Other Special Microscopy Tests 74
Microbiological Plating and Media Management 77
Sampling from Tanks and Zwickel Maintenance 78
Pipetting and Other Challenges with Microbiological Plating 79
Chemistry Tests 82
Measuring Extract and Alcohol Background 83
Measuring Extract on the Brewdeck 84

  Spectral Analysis 85
  Measuring pH 86
 Sensory Analysis 86
  Taste Panel Selection 87
  Criteria for Go/No-Go 89
 Packaging Tests 90
  Carbon Dioxide Checks 91
  Measuring Dissolved Oxygen 94

7. Government Affairs 97
 Food Safety and Risk Assessment 97
  Food Plant Registration 98
  GMPs – A Foundational Requirement 98
  Fill Level and Alcohol Monitoring 100

8. Pulling It All Together – Assessment Time 103
 Types of Audits 104
 Conducting a Quality System Audit 105
 The Quality System Audit in Three Parts 106
  Process Assessment 106
  Department Assessment 108
  Product Assessment 108

Appendix A Small Brewery Quality Manual Example 111
Appendix B Quality Control and Assurance Plans 121
Appendix C HACCP Risk Assessment and Critical Control Points 131
Appendix D Failure Modes Effects Analysis (FMEA) Table Example 135
Appendix E HACCP Process Map with CCPs 137
Appendix F Standard Operating Procedure (SOP) Example 139
Appendix G Quality Inspections for Maintenance 141
Appendix H Good Manufacturing Practices (GMP) Policy Example 143
Appendix I New Product Quality Control Plan Example 149
Appendix J General Audit Report 153

Glossary 159
Resources 161
Bibliography 165
Index 169

ACKNOWLEDGMENTS

Special thanks to those whose guidance and technical eye made this book reach a broader audience and helped highlight critical elements of quality management. Specifically, I'd like to thank Chris Swersey and Chuck Skypeck from the Brewers Association for their guidance and work ensuring our alignment with the Brewers Association Quality Subcommittee. Thank you to Jason Perkins, Brewmaster at Allagash Brewing Company and Brewers Association Quality Subcommittee Chair, for his review and for being willing to write the preface. Dana Sedin, Brewing Chemist/Sensory and Chemistry Lab Manager at New Belgium, did a great job in his work as technical editor.

It is a special honor to have the foreword of this book written by Ken Grossman, a leader of our industry and advocate for quality beer since he started Sierra Nevada Brewing Company. Thanks, Ken, for your leading comments and for continuing to set the quality bar high.

Thanks also to my previous co-workers, peers, and mentors at the breweries I have worked in. Your passion, endurance, and joy for the trade made me want to learn my craft in service to you. In addition, I want to thank the dedicated staff at the Brewers Association for the honor of being able to write such an important book, and to Kristi Switzer for her editing guidance and endurance.

Finally, thanks to my husband, Noah, and our children, Lucas and Nils, for their tolerance, patience, and love as I took the time away from my family to write this book.

FOREWORD

I have never heard a brewer or brewery owner say that they don't want to make the finest quality beer. I truly hope it's what we all aspire to when we walk through the doors of our breweries every day. The insistence on quality is the tenet that has in large part driven the current and truly amazing success of the craft brewing segment.

Whether you are new to this industry or have spent long a career making beer and learning from the mistakes of those who came before, quality can prove elusive. The adage that brewing is part art and part science is often repeated and certainly holds a lot of validity. Rest assured, embracing good, well-reasoned quality control measures should in no way impair the art of brewing. Innovation has helped drive the incredible growth of craft brewing and it would be a mistake to see it as being at odds with quality. We should acknowledge the symbiotic relationship between the two—the part of innovation that's exciting and invigorating needs the balance provided by discipline of quality. The pursuit of innovation and quality should be every brewer's passion.

Making great beer is hard. Making the same great beer every day is harder. Ensuring that your great beer is still great after packaging is harder still. And the hardest thing of all is ensuring that every customer gets the great beer you brewed whether they drink it in your pub or another establishment or from a glass, bottle, or can. I'm certainly not trying to discourage anyone who recently entered the brewing industry, but after 35 years I'm still learning how hard it is to produce quality beer day in and day out. As Mary very rightly points out, "The practice of quality management takes a lifetime of practice to master."

I tend to believe that all brewers have the same high regard for quality, but I often see a disconnect when it comes to putting those beliefs into action. Simply put—some brewers talk the talk, but don't walk the walk. Fortunately, many components of a thorough quality system don't have big upfront price tags, but opting out of them can prove costly when you have a quality problem. That cost comes not only in dollars, but in losing the faith and support of wholesalers, retailers, and consumers, and it certainly puts your brewery's reputation at stake with incalculable marketplace consequences.

I hate to say it, and I certainly hate to do it, but sometimes dumping beer is necessary and can even be a good thing. I dumped nearly 10 batches of Sierra Nevada Pale Ale that I intended to sell as the first batch before I finally hit upon the formula to great, consistent beer. Making the decision to dump that beer was agonizing because each discarded batch only added to our financial woes. But I knew it had to be done. I wouldn't stake my name or that of my brewery on less than great beer. When a brewery owner or head brewer knowingly releases sub-par beer, it sends a terrible message to everyone involved in ensuring quality (and that should be everyone at your brewery). Dumping a batch of beer for which you know every single line item cost can be painful, but if that beer's quality is compromised, getting rid of it is not only the right thing to do, it helps foster a culture of quality in your company.

Having said that, no brewery produces fault-free beer 100 percent of the time. Analyzing every specification, dissecting our beer down to the most granular level won't ensure that we achieve perfection—both in the science and in the art of brewing there are a multitude of factors that could always be more tightly controlled or improved upon. We can, and should at a minimum, analyze for parameters such as color, clarity, carbonation, alcohol content, head retention, shelf stability, and bitterness—it's a lot of analysis for every beer your brewery produces, but as our consumers become more educated and the marketplace becomes more competitive having really great consistent beer is table stakes to stay relevant as a brewer. Fortunately, achieving perfection isn't the point of this book or the point of a robust quality system. Breweries large and small, old and new, can always make better beer and that should be every brewer's goal. Mary lays out many tools to help in this quest, but be prepared for a journey that has no final destination—you can always travel further down this path.

In the early days, craft brewers had to develop their own quality management practices and systems without much to go on. The big brewers provided good examples, but their equipment and labor resources were out of the reach of most small brewers. Fortunately, those of us who got our start by homebrewing—and managed to survive—generally developed a good understanding of the quality pitfalls that plague brewers and were able to build on that foundation. As Mary explains, quality management systems are scalable. Some of the equipment and tools that a 100,000-barrel brewery uses to ensure quality will most likely be out of reach of a brewery that produces 2,000 barrels. However, the basic elements of a robust quality system remain the same: monitoring, measuring, and reporting built on a solid foundation of training and education. Regardless of size or budget, with proper direction and focus you can develop a great quality culture and plan.

A good quality management program is more than just lab analysis. It encompasses a Good Manufacturing Plan (GMP), HACCP (Hazard Analysis and Critical Control Points), documentation, practice, and attitude. As I've said, the latter may be the most important element in determining the success of even the best written quality management program. It is incumbent on brewers and brewery owners to create an environment where employees feel comfortable speaking up about quality concerns and confident that a clear line of decision-making authority will address those concerns. A quality culture requires cross-departmental cooperation and communication. A good quality management system follows the entire brewing process, from raw material selection to packaging materials. In a large brewery, different people may be responsible for each step in the process, but one of the most important things they can do is work together to ensure that the product they produce arrives at its downstream destination in specification.

In my opinion, the most basic, yet frequently overlooked or underestimated, component of beer quality is cleanliness. It's a make or break proposition—you can't have a dirty brewery and consistently make clean beer. When Sierra Nevada was starting out and I couldn't afford many employees, I spent many hours of very long days cleaning the brewery—sanitizing equipment, mopping the floor, clearing debris from production areas. As the business and brewery grew, I was able to hire sanitation employees to keep the brewery clean. From the floor drains to light fixtures and from tank tops to tank outlet valves, nothing is overlooked. We've found that when people work in Sanitation they develop a great attitude toward and habits around cleanliness that serves them well as they move into production departments. It's no coincidence that we call Sanitation our "farm team"—it has served as a starting point for many Sierra Nevada employees.

After cleanliness, the next crucial step in developing a good quality system is measurement—if you aren't measuring, you have no way to know if your beer is on target and no way to track improvement. I once spoke to a mid-sized brewer whose beers were often oxidized in the trade. He told me that his O_2 meter had been broken for several years, but his packaged O_2 levels were fine when he last measured them. I was both shocked and amused. He might as well have said that his car's speedometer was broken, but he was confident that he always drove the speed limit because he had been the last time he'd been able to check.

We know that oxygen pickup after fermentation is bad for almost all beer styles—it does nothing but degrade beer stability and denigrate the flavors we work so hard to develop. Oxygen should always be at the forefront of your and your brewer's minds. That may sound funny coming from the CEO of an established brewery, but if you ask any of my production team, they'll tell you that I always harp about oxygen pickup. These days it can be measured relatively inexpensively and it should be one of the basic quality indicators and checks that brewers of every size embrace. Still, it's not that simple. Oxygen in beer is transient rather than static. It is very good at what it does and ruthlessly oxidizes beer. After the yeast has taken up the oxygen it needs in the pitched wort and is done with its fermentation cycle, added oxygen isn't your beer's friend. However, if you don't look for ingress along the entire process, it may well be gone when you test it at bottling (hopefully you're testing at bottling).

Chasing quality and demanding improvement is truly a philosophy. It's not a box to check off a form or a test result to enter into a spreadsheet or database. Quality is a mindset and a way of working, not a number. Although this book is intended for, and will probably find its primary audience among, production-oriented brewery employees, brewery owners have to spearhead their company's focus on and commitment to quality, whether they are brewers, passive investors, or financial groups. Any brewery owner who wants employees to advocate for quality has to do the same.

That said, breweries are capital-intensive ventures and the demands for an ongoing supply of cash are numerous. I won't tell another brewery how to allocate its resources other than to admonish those who do not continue to invest in building their quality programs as they attempt to grow their companies. Our collective success is built on quality. I hope that both established and aspiring brewers can learn and benefit from many of the fundamentals that Mary details in this book.

Ken Grossman
Sierra Nevada Brewing Co.

PREFACE

As professional brewers, we are living in extremely exciting times. One could easily argue that there has not been a time or place in the history of the world that is as exciting to be brewing beer as in this decade, in this country. Craft brewers have repeatedly showcased their ability to produce unique flavor experiences; every day more and more consumers experience them for the first time. We are very fortunate to have a consumer base that demands these flavors and has tremendous confidence in our beer.

With this excitement and growth comes near endless opportunity. New breweries are opening with remarkable frequency and existing breweries are growing at a breakneck pace. But all this growth and opportunity carries with it tremendous responsibility. We owe it to our consumers and the general public to produce beer that is safe to consume and is of the highest quality.

The level of responsibility is the same whether you are a one-person operation or a brewery of several hundred employees. Big or small, all brewers share the same burdens for responsible beer production. Before a brewery opens its doors, or the first shovel hits the dirt on a major expansion, the foundations of a quality program must be a major part of the planning process. All the wonderful opportunities that currently surround us will disappear if consumers begin to lose confidence in our beers.

The pursuit of quality is a life-long quest. As Mary points out in this book, it is one that no one will ever master, a journey that never ends.

For those starting a brewery without any brewing experience, or brewers taking on a major expansion, incorporating a thorough quality program during the planning process can be really intimidating. Especially for a brewer without a scientific background, some of the concepts of Quality Assurance can seem daunting.

Don't let this affect your focus. Start small and tackle the low-hanging fruit. Climb the Quality Pyramid one layer at a time. As Mary eloquently points out in this book, achieving quality is like "eating an elephant," you need to do it one piece at a time. While you may not have the financial or logistical resources to implement every quality procedure into your program, the general tenets of a great quality program are applicable at every level. Brewers are creative and ingenious people—tap into that and find methods that work at your size and resource level.

Most importantly, start a strong quality-focused culture early—a culture that incorporates a focus on quality throughout the brewery. All staff are essential in monitoring quality, from the person who accepts the grain delivery all the way to the sales representative in the field monitoring the quality at retail. In many cases, it may actually be the same person.

If you wait to start a quality program, establishing

it as part of the culture after the fact—after you get through the first year or once the latest expansion is done—is all the more difficult. Like a large ship in the ocean, it is far easier to start in the right direction than it is to turn around once it's set on a course in the wrong direction.

Mary Pellettieri has created a tremendous resource for brewers; this book is packed full with quality concepts. Every one of these tenets is applicable to all brewers, regardless of size. As you read this book, you may come to sections where you think, "I'm too small to do that" or "I don't have the resources to implement that." When this happens, stop yourself and think twice.

Don't be intimidated. There may be a way for you to create a procedure for your current size and situation. At the very least, every brewer should be educated about every concept in this book and put thought in to how these tools can help, at least some day.

In the end, it is up to the individual brewery to set its quality priorities. The concepts and tools provided in this book offer tremendous guidance along the path to shaping a superior quality program.

Jason Perkins, Brewmaster
Allagash Brewing
Chairman, Brewers Association
Quality Subcommittee

INTRODUCTION

The practice of determining if a product such as beer is acceptable enough to send to the customer is the practice of quality management. This is easily said, but this one sentence requires a lifetime of practice to master. I am lucky enough to have been responsible for quality management in a brewery environment for most of my career. Ranging from a growing small craft brewery to facilitating quality in a large brewery, I have tended to quality since the early 1990s. Not only did I learn and grow as a brewer, but I also learned about the world of operational excellence that comes from the study and application of quality management.

This book was a labor of love, borne out of my interest to see beer quality improve with the explosion of breweries, style advancements, and innovations. This interest has been deeply embedded in me from my personal experience. In the early 1990s, small brewers had to build and rapidly evolve quality plans so we didn't lose to the "competition" (the very large breweries, and imported beer of the time). We had a learning curve, certainly. Many of us didn't have the years of quality management experience under our belts that larger breweries did. Terms such as Six Sigma and Hazard Analysis & Critical Control Points (HACCP) were new to us. Still, there was a "team" aspect to being a part of the craft beer community. We knew we were starting a new chapter in the American brewing saga, one in which innovation was leading the charge. No one wanted to be the brewery that allowed innovation to run too far ahead of quality. We didn't want to "screw it up."

This ethos remains true today in both small and large breweries, both challenged to innovate quickly, but it is especially true for craft brewers who have grown a "brand" solely on the fact that they changed the paradigm of what quality beer can be. Fully knowing the cost of poor quality may result in the loss of the customer's confidence. Ironically, innovation can challenge any good quality management system to the point of failing. Losing this push-pull game between quality and innovation is not an option. *Innovation brings customers to craft beer; quality keeps them.*

The biggest challenge to smaller brewers, just like I experienced in the early days of my career, is "eating the elephant" one bite at a time, continually adding to the quality system, and keeping up with growth and new products. The required work to establish a robust quality management system is extensive, even if growth is not a factor. From writing policy, procedures, and specifications to establishing protocols, corrective actions, and improvements, it can feel overwhelming, but if tackled one task at a time it is very doable—and vital to success! Remember that although the standard set of rules in quality management are written, codified, and even well understood, they are rarely static. Instead, the rules (the policy and practices of quality), are slowly built upon a foundation and desire for excellence and evolve with growth and innovation. Quality management practices build on that foundation, growing with the risks and the complexity of the business. And, most importantly, the practices have to evolve continuously so no one initiative, such as growth or innovation, runs faster.

The brewing industry is rife with stories of quality being trumped by growth or innovation, only to result in the loss or significant shrinkage of the brand. A famous brewery mentioned in Chapter One of this book is a good example of this. Craft beer, as a segment of the highly evolved and long-standing commercial brewing industry, brought innovations of reviving older brewing techniques, yeast strains from across the globe, and using materials in ways never before conceived. Craft brewers were free from the constraints that beer had to be one flavor, one color, one yeast. Making these new styles of beer required a sense of beer quality and technical knowledge. However, craft brewers have always risked much in the desire to innovate and create something "new"—this can slowly erode or outpace a quality system without dedication to quality. This was the risk craft beer faced at the start, and continues to face.

So how to eat the elephant? One bite at a time, and then don't stop (those of you with some experience likely realize this.) This book is designed to provide a broad overview of quality management from a brewer's lens. It was also written to apply the practice as a separate but much-needed co-partner to technical brewing. Though smaller brewers may find much of this new, and a bit overwhelming, taking the advice one step at a time is recommended. For example, don't start a continuous improvement team before you have the basics of quality management knocked out first. We will discuss these, but writing Good Manufacturing Practices (GMPs), establishing policy, setting specifications, writing up procedures, practicing governance, defining your risk assessment, etc. must come first. Use this text to develop your first five objectives this year for what you need to do to start or improve your quality management system. Then pick it up again as you plan for the future every year, build new objectives, and don't stop. If the quality management system keeps up, evolves, and stays ahead of the pace of growth and risk of innovation, the brewery has the best opportunity to compete and thrive in the growing and exciting landscape of beer making.

ONE

DEFINING QUALITY IN A BREWERY

The expected outcome of a quality program, in any type of manufacturing environment, is to produce high quality products in a consistent manner. For a brewery, that translates to making quality beer all the time. This job falls to all workers in the brewery, but the quality system—the management of the resources to make consistent beer all the time—usually resides with one person. In a very small brewery it may be the owner, or the head brewer, that has to wear this hat. Larger facilities usually have a position dedicated to quality. Regardless of who is wearing the quality hat, these folks frequently hear, "What a fun job! You get to taste beer all day!" Working in a brewery can be fun; however, leading the continual quest for quality is a demanding job. Tackling an overall quality plan can feel a bit overwhelming, overstructured, rigid, and sometimes undervalued…until something goes wrong. That is when all the planning and testing in place pays off.

A quality manager in a food plant, where microbiological food safety issues are closely tied to the quality of the product, requires clear delineation of what is *good* and *poor* quality. They may pose the question, "Will releasing this product hurt our customer or make them ill (e.g., due to microbiological pathogens)?" An ice cream facility, for example, could release a batch of ice cream contaminated with pathogens that could sicken customers.

In a brewery, on the other hand, the quality of the product is primarily defined by standards of what constitutes "good" flavor, color, foam, shelf life, customer expectations, etc. There the question may be, "Will distributing this batch of beer hurt our customer's *response* to our beer?" We cannot hurt our customers by making poor quality beer (in terms of microbiology since pathogens cannot survive in beer). However, brewers can hurt their business reputation and the reputation of *all* craft beer. With a lower level of risk of consumer harm, the role of quality management demands in the brewing industry differ from food products and the system may easily become muddled between decision makers in brewing operations and the quality lab staff. For reasons unique to beer brewing's history, management gaps remain in regard to defining quality, determining how to achieve it, and who is responsible in the brewery for maintaining product quality. When these questions are left unanswered, the brewery may indeed suffer catastrophic quality issues.

This chapter explores the history of both manufacturing quality and brewing quality. These two disciplines are very much related, but there hasn't been a full analysis of the interplay over the last 50 years of modern industrial history. Both subjects require a large set of knowledge and skills. The discipline of manufacturing quality requires an understanding of statistics, risk analysis, communications, and change management. The discipline of brewing quality requires understanding of the brewing process, microbiology, chemistry, sensory analysis, and understanding variation, statistics, and measurement. Quality managers in breweries must master both manufacturing and brewing quality disciplines to be effective. The person who wears the quality manager's hat must bridge any gap in knowledge with solid training, hands-on experience, and a lot of reading.

If the strength of the quality manager is in brewing knowledge, which many times is the case in small and growing breweries, there is a tendency to get lost in the data-rich environment and neglect to ask the big questions that well-trained quality managers would first ask, such as, "What are the key quality criteria for our brand?" And, "Who has what responsibilities toward maintaining our product quality?" These are truly quality management questions, and this is a good starting place for developing a plan. As breweries grow in volume, change and add products, and maybe even change leadership, it becomes increasingly important to take a broad look at *what* you define as quality in your beer, *how* you do it, and *who* has what responsibilities. Quality science, like brewing science, has continuously evolved; and breweries must adjust their quality management style to grow with the brewery.

There are several reasons achieving consistency and excellence in product quality can be a challenge unique to the brewing industry. First, define specifications for the ideal batch of beer. This alone is not hard to achieve, but the specification must be bolstered by a management system of policies, procedures, and human resource practices that allow employees to correct a process. With so many data points to measure in a brewery, this becomes a bit ghastly. Most importantly, there must be a functional role or advocate leading the creation of the culture of quality in the brewery, and it must be emphasized from the top down to be successful.

"What are the key quality criteria for our brand?" And, "Who has what responsibilities toward maintaining our product quality?" These are truly quality management questions, and this is a good starting place for developing a plan.

The advocate may be the brewmaster, the CEO, the quality manager, or all of the above. In a highly technical field such as brewing science, it is sometimes easier to delegate the leadership to one person in the brewery. The quality advocate has a specific and crucial role to play—to ensure *everyone* shares the responsibility of producing quality results. If a brewery establishes a quality-focused culture throughout the company, it will always produce a superior product, despite lacking any of the other requirements. This is one of the fundamental tenets of developing an effective quality program; without strong leadership to create an effective focus on quality, established quality systems are for naught.

In a small brewery it can be easy to push off the structure of a quality system, especially during start-up mode. There are plenty of duties and issues to resolve, and the last thing on everyone's mind is to codify specifications and requirements for releasing products. However, the more structure put in place early on, the more smoothly the brewery will function as it grows. We will discuss structure more in this book to make it easier to implement in stages.

The Changing Definition of Quality Management
Let's briefly look at history in both quality and brewing to help divulge some of the unique challenges the brewing industry has in terms of quality management. Quality management, as a field of study, has been evolving throughout modern day manufacturing history. There has been an evolution of quality management from the early days of Total Quality Management (TQM) in the 1970s to today's focus on Six Sigma. The brewing industry was not fully engaged with the field of quality management in the early heyday of the quality management evolution in the 1950s and '60s (as we will expand upon in a bit). This was left more to the automotive industry.

However, as each industry became more engaged in the study of quality management, the overall product quality improved (Vrellas, 2015). The automotive industry learned the hard way that quality as a system and a field of study is important to prevent failures, and because they implemented the systems discussed in this book, they improved steadily the expectations of consumers during this evolution. For example, in the 1960s, cars failing in any way (e.g., poor materials on the interior, knobs breaking off, etc.) during the first five years of use was not considered a failure of manufacturing quality. Today, however, cars have a manufacturer's warranty of at least five years. In other words, with the implementation of quality systems and, effectively, continuous improvement, the expectation of quality in automobiles continues to change as consumers raise the bar year after year.

Quality management is not just a set of criteria that define *good* beer, but it is also a system of policies, procedures, specifications, and empowering employees at all levels to correct a process.

The perception of value or quality in beer is no different. With every new generation, beer quality is evolving. These changing standards, along with the concurrent growth of the industry, make bringing in broad quality management philosophies of business excellence too much for some. In fact, breweries are still playing catch-up with other industries in their development of quality management practices (Vrellas, 2015). To understand more deeply where the divide started, it's important to review what was happening in the brewing industry at the same time the manufacturing industry began its immersion into quality management philosophies.

The Gurus of Quality
This discussion of quality management as a science and study can be credited to several theorists who challenged our thoughts on what makes not only a good quality product, but also a good quality practice. W. Edwards Deming, Joseph Juran, and Philip Crosby are often referred to as the gurus of quality. There are others, but these gentlemen were instrumental in shaping what good quality management is and how to achieve it. And because they will be referred to frequently in this book, a little background on these individuals is warranted.

Deming, a mathematician at the US Bureau of Census, used statistics and sampling methods that greatly improved productivity in the 1940s census. He brought a discipline of measurement to Japanese manufacturing after the war and assisted that country in its rebuilding efforts. Deming is credited not only for unifying leaders of the brand names we know, such as Sony, Toyota, Nissan, etc., in the use of advanced statistical quality control techniques, but also for helping these leaders understand the need to bring statistical reasoning to the front line. Deming believed the focus of management should be on reduction in variation. In the brewing environment, for example, this would translate to continuously reducing the variation in dissolved oxygen. Deming believed quality and productivity always improved with focus on variation reduction, and this philosophy is still incorporated in today's modern manufacturing plants, namely in the automotive industry (Bank, 1992, 60–82).

During the 1940s and '50s breweries started to expand, centralize services, and bring in a statistically minded methodology for making decisions, incorporating Deming's philosophies. Mortimer Brenner's 1953 paper "Some Thoughts on Quality Control" in the archives of the Master Brewers Association of the Americas (MBAA) *Technical Quarterly* gives extraordinary detail on how to establish statistical process control (SPC) in a brewery. Brenner was right in line with the latest thoughts when he implored the industry to take up SPC, illustrating the influence Deming's philosophy had on the brewing industry. The practices of reduction of variation continue in breweries of varying size today.

> *"One may set high goals for quality factors, but the progress toward the goal should be watched with the help of statistical aids. Control charts will be very helpful in deciding whether results are out of normal limits because of assignable causes or whether an operation has reached the best performance, which may be expected, short of a significant change in processing equipment or procedures."* (Brenner, 1996, 193–199)

Where we begin to see some divide in the evolution of quality management and brewery practices was with Juran in the 1960s. Joseph M. Juran worked for the manufacturing industry, and later as the Head of the Department of Industrial Engineering at New York University. He wrote the first copy of the *Quality Control Handbook* in 1951, published by the American Society of Quality (ASQ). The sixth edition, published in 2010, is nearly triple the size of the first edition. Juran's quintessential book was not so much a technical reference as it was a manual for the human side of management. He showed leaders how to manage employees with the goal of quality. His 1964 book, *Managerial Breakthrough*, addressed resistance to change and how to create an infrastructure of improvement using teams. His "trilogy" of planning, control, and improvement is still relevant to brewers today, though its practice did not take root at the same pace it did in other industries. We will explore Juran's definition of quality as "Fitness for Use" later in this chapter (Westcott, 2005, vxii).

The last quality guru who helped shape the definition of quality management in manufacturing, Philip B. Crosby, introduced the concepts of prevention, not just inspection, and Total Quality Management (TQM) in the 1970s and '80s (Westcott, 2005, 49–52). Crosby worked for International Telephone and Telegraph for several years before starting his own quality consulting firm in the '80s. He defined quality as "free from defect" and brought to the attention of manufacturers that quality is NOT just the responsibility of the quality department, but of all the members of an organization (Bank, 1992, 60–82).

TQM was a trendy management philosophy throughout the '90s. It was a way for organizations to incorporate the philosophies of Deming and Juran. It standardized tool usage, such as control charts and statistical methods, by training management on the fundamentals of statistical inference and data-driven decision making, as well as organizing process improvement and problem solving. TQM has since been replaced by other standardized quality improvement methodologies such as ISO-9000 (a quality standard), Six Sigma, and Lean Management. These programs help standardize the

---

Quality Certification Types

ISO-9001
A quality management certification program. International Organization of Standardization (ISO) supplies a set of standards to comply with and be audited against. Company certification.

Six Sigma
Six Sigma organizes a continuous improvement program and reduces process variation, thus improving product quality. Individual certifications follow belt levels (from green belt to black belt).

Lean Management
Lean management is another type of continuous improvement program and its techniques are usually used in conjunction with Six Sigma. There are individual certifications in lean management.

HACCP
Hazard Analysis and Critical Control Point is a food safety and risk management tool that requires training and certification of individuals. A plant certification is conducted by an audit performed by an outside company.

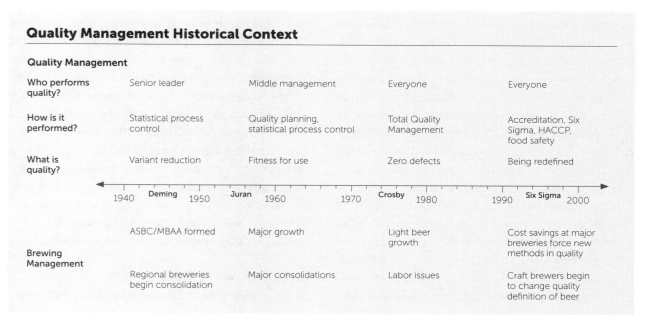

Figure 1.1: Timeline of the development of quality management practices in the brewing industry

who, what, and how of a quality system *and* the process of improvement. They are sporadically embraced by the brewing industry (see side bar). It was Crosby who brought a disciplined approach to quality system standardization and process improvement. Crosby's philosophies strike not only what quality is, but also how things are done. (Bank, 1992, 60–82)

There has been tremendous change and evolution throughout the last 50 years of not only what quality is, but also who is responsible and how they achieve it. The development started with Deming introducing SPC, then Juran introducing the management role and front line quality, and lastly with Crosby introducing a discipline to problem solving. Let's sum these up again as we will take them on one step at a time in this chapter and review how these relate to brewing. Figure 1.1 is a good visual of the items and their change across time.

Quality management components:
1. What: The criteria of what "good" looks like.
2. How: The process of developing quality criteria for a product, and then applying that criteria to the operation.
3. Who: The people who dedicate themselves to quality, and who hold other roles and responsibilities throughout the brewery.

WHAT IS QUALITY?

"A company will know it is producing high quality products if those products satisfy the demands of the marketplace." (Ryan, 2000, 3)

In the fall of 2014, the Brewers Association (BA) Technical Committee formed the Quality Subcommittee, which sought to identify a general definition for BA members on quality. After much deliberation they agreed that "Quality is a beer that is responsibly produced using wholesome ingredients, consistent brewing techniques, and good manufacturing practices, which exhibits flavor characteristics that are consistently aligned with both the brewer's and beer drinker's expectations." Quality beer is the responsibility of all brewers and is necessary to ensure safe products.

Breaking this quote down, "quality beer" is made in a certain way, is safe to consume, and there is a system to monitor the process from raw material selection through brewing and fermentation, from packaging to serving the end consumer. This template definition incorporates traditional views of defining quality.

Defining Quality as "Free from Defect"
A traditional way to define quality is "free from defect."

As such, your quality goals are to prevent off-flavors and meet government requirements. Brewing science schools train students to understand how defects can get into beer and how to prevent them. They are also trained in how to ensure food-safe beer. United States laws require tight controls on weights and measures (bottle volume), alcohol percent (if the bottle is labeled), and sulfite concentrations, and that beer is made in a food-safe manner. That is it. Quality as "free from defect" is the baseline.

Outside cultural pressures or even the government can manipulate quality as a standard of excellence and "free from defect." A great example of changing quality standards due to regulation is when German brewers were subjected to the Reinheitsgebot in 1516. It forbade using anything but barley malt, hops, and water to make beer. At the time, not only were oats and wheat interchanged when malt was in short supply, but hops were replaced with other bitter herbs on occasion (Corran, 1975, 261, 279). The "purity law" applied throughout Germany beginning in 1906 and this, no doubt, forced a certain flavor profile because of the specific raw materials (Kunze, 1999, 20). The Reinheitsgebot standardized what was "free from defect" in beer, and is still considered an early example of a quality standard that had a significant impact on beer.

The criteria of a "defective beer" have also changed in North America, especially over the last 200 years. In the 1800s, brewers struggled with sour, spoiled beer in the market, and devised yeast handling procedures and temperature control. After prohibition, consumers demanded light lagers—described as lightly alcoholic, low bodied, and bitter—and they became all the rage (Barron, 1975).

Starting in the 1990s with the growth of craft brewers making ales and hop-forward beers, the standard of quality as "free from defect" is changing again. The standard of using light lager characteristics to define "good" beer is being replaced by the diverse and inventive styles of craft brewers, leading to more complex challenges in quality. There are now many different styles of beers, a wider variety of raw materials, the introduction of new brewing processes, and historical processes being revived. The result is a multitude of flavors and therefore new definitions of what quality beer is. Imagine telling a new-to-craft-beer customer that "sour flavor is ok in this beer, but not that one." That is confusing for the customer and difficult for the brewer and marketer because it requires time and education of the consumer. A customer may often need to be educated on style characteristics in order to know that sour can be a desired trait in certain beers but an off-flavor in others. Some of the flavors that were once considered "poor quality" are now coming back in favor with new and evolving styles. Today's brewers have a difficult task as they must define again and again for the customer what good quality is, and educated consumers will understand which flavors they enjoy or dislike, adding to their own knowledge of quality.

Defining Quality as "Fitness for Use"
If the answer to the question "What is quality?" results in "fitness for use," this becomes a bit more difficult to translate when relating to beer quality. "Fitness for use," or what Guru Crosby introduced, assumes the customer helps define what a good quality product is. Brewers, in their effort to meet customer driven quality, may measure many parameters throughout the process and, unfortunately, make certain specifications tighter than what the customer would ever pick up on.

The parameters commonly used in beer manufacturing that define "fitness for use" are color, bitterness, alcohol, and foam, which reflect what customers most value in beer (Bamforth, 1985, 154–160). Added to this list could be beer clarity and flavor profile. If all these parameters fall within the defined style or occasion for which the beer was designed, then that beer is one of quality as defined by "fitness for use." Of course, if these parameters are exceptionally variable from batch to batch, and that variation exceeds what is deemed acceptable by the customer, then quality is being compromised. This variation is highly dependent on what the customer sees as useful. A beer that varies in degree of haze or sediment by a wide margin may not be acceptable to the brewer, or even for the beer style as defined, but it may be perfectly fine for the customer. Alternately, the same customer may insist on a flavor characteristic that is very difficult to consistently replicate. If that customer experiences a swing in the flavor they like or dislike too many times in one brand, they'll go elsewhere. This is a special challenge for the brewing industry.

Sometimes controlling quality criteria too tightly can be the challenge. Overspending to control a quality criterion that is considered important for excellence but underspending to control a criterion the customer values most can easily result in loss of business. Color, bitterness, turbidity, and even flavor can all vary according to what the customer requires. Yet these parameters are also kept under strict control—sometimes too strict. The quality management philosophy that Crosby related was to drive down variation parameters so there is never a risk of releasing beer that doesn't meet the customer's basic needs. Defining those basic needs is your starting point for all beers. This may result in parameters that have different levels of variation. For example, a hop-forward beer may not need the same level of bitterness control as a less bitter beer. If the brewery struggles with controlling one of the primary parameters in beer that is harder to control, it is time to look at what the customer's needs are, assuming you agree that quality is "fitness for use." Using customer-defined specifications to guide your process control may save a lot of time, money, and effort in the end.

Defining customer-derived quality specifications is easier said than done. When developing a new beer, the key is defining what criteria *and* variations are acceptable, and aligning those to the leadership, and sales and marketing, expectations. For example, if a beer is clear, but a slight haze or sediment would be acceptable to the customer, it is highly recommended the specifications state this. As that brand expands in geography, it is possible the risk of a slight haze may also increase. If it isn't stated at the conception of the brand that a slight haze is acceptable, and the brewery begins to get complaints from the sales team that the beer is hazy, the brewer may add a chill-haze compound of some sort during processing to keep what is an unstated quality parameter intact and thereby increase costs. Therefore, the benefit of articulating your beer in a "fitness for use" manner is that it forces brewery leadership, and sales and marketing teams, to identify and align to what is critical to the customer. Then the brewing process stays in control of what is critical and nothing else. If everyone agrees on the parameters that must remain within certain customer-derived limits, then deciding what to do with beer or products outside those limits is less of a discussion and more of an action.

Defining Quality as Esoteric
What about the *art* of brewing? For some, just adhering to customer-derived limits of alcohol, bitterness, and color lacks the essence of the experience that particular beer should evoke. How do we address this when defining what quality is? Those of you who are unsatisfied with a clinical definition of every parameter as a way to evoke quality should not limit your quality definition to "fitness for use" criteria only. Quality in your breweries may extend to defining "brand values"—that is, how that brand differentiates itself to the customer. For example, you may see your beer as not just a vehicle to deliver alcohol, color, and foam in a specific form, but instead it is a sensory enhancement to a meal, or to a conversation. If that is the case, take the time to define this value as a criterion, such as "drinkability" or "complementary to food," and write it down as a means to help management make better decisions. It is perfectly acceptable to articulate values to a quality manual in terms of criteria as part of how your brewery defines brand quality. As a brewery scales a brand in size from a small single batch to multiple batches, to blending and large volumes, having a firm hold on the values and quality characteristics that customers most desire and associate with the brand is ultimately why quality criteria are needed.

Traditional Foods and Quality

What makes smaller batches and traditionally crafted foods and beverages unique? They have the benefit of adding a handcrafted value to the product that an industrialized equivalent cannot replicate. For example, the sour note in beer achieved by months, if not years, in a barrel has a cost of time, space, and labor. It may require some consumer education to understand why it is important, and how it adds to the beer complexity. There is inherent value in the time and art of the brewing process to get that particular flavor and educate the consumer to appreciate it, and, therefore, pay more for it.

HOW TO ACHIEVE QUALITY – THE QUALITY SYSTEM

The quality system is a term used to define the overarching system of process, policy, and record keeping that maintains control of the desired quality output. In brewing, quality should be found at every step of the process. The process of brewing requires a series of interlinked, highly technical steps that take years to master. Each function, from wort boiling to fermenting to finishing to packaging, has to work together in a chain. At the connections of the chain links are the quality checks. These checks measure if the process was in control. For example, after mash mixing is complete, the brewer checks to see if starch conversion is complete via an iodine check. If this check passes, the mash moves into the lauter tun.

> **Process Input Measurements vs. Process Output Measurements**
>
> If a quality check happens after a process has completed its cycle, it is called a "process output check" or "process output measurement" (POM). The iodine check is an example of a POM. In addition to the POM checks, the checkpoints that monitor a process during the operation are called "process input measurements" (PIMs). These checks may be done manually or be automated and controlled by the process logic controller (PLC).
>
> In the case of the mash mixer, the PIMs may be temperature control and time. By controlling the PIMs, process controllers prevent the production of defective product in the first place, providing it is properly programmed and monitored. As the ability to measure both process outputs and inputs improves so will the overall product quality, because monitoring and adjusting before a process is out of control manages product variation.

Excellent communication and documentation is required for this check system to work. The record of each step's key process indicators (outputs or inputs), process anomalies, and what was done to correct them, is highly valued when troubleshooting. If a mash is not moved forward because it fails an iodine check, the brewers can look at the data from the process controller (if documented) and determine if a switch or valve did not trigger at the right time and allow the mash to properly heat. All of the information can be summed in a "scorecard" or some sort of communication vehicle. This communication is important because it directs management's attention to process control, and therefore quality control. The entire picture from training on the process, measuring and documenting each step, maintaining and monitoring records for trends, coordinating the communications, and reviewing the data for improvements is effectively a "quality system." All of these components need to be championed and managed.

The entire picture from training on the process, measuring and documenting each step, maintaining and monitoring records for trends, coordinating the communications, and reviewing the data for improvements is effectively a "quality system." All of these components need to be championed and managed.

The quality system in any environment should look somewhat similar. Not unlike accounting or standard financial reporting, quality principles, reporting, and processes are very applicable to every business and ultimately drive performance. This seems intuitive at first blush, but in practice it can get complicated. Many factors unique to the brewing industry, such as the technology and the required base knowledge and skills, add to the complexity of the overall quality system. Interestingly, quality management related skills can be the strongest in highly technical industries. As early adopters of the philosophies of Deming and Juran, Toyota, Motorola, and General Motors are a few of the companies that stand out as leaders in the area of quality. Most readers will associate traditional quality systems as applied to automotive manufacturing. These companies benefitted from adopting and evolving quality principles such as Statistical Process Control (SPC), managing processes using good data analysis, and as technology advanced they even used statistics. Summarizing all of their data in scorecards, they effectively simplified and organized their operations to continuously improve.

The brewing industry was not totally naïve to the foundations of establishing a good quality system when Deming and Juran started to share their philosophies of managing by data analysis in the 1950s and '60s. The processes to make beer have been developed and improved upon for the last two centuries because of good data processing and statistical controls. Many in the beer industry recognize the Student's T-test, a statistical test to determine if two populations have a similar mean that was developed by a chemist at Guinness & Co. in 1908 (Raju, 2005, 732–5). Data gathering and reporting is a discipline taught with rigor in brewing schools, and brewing chemists have been an active part of brewery personnel since the 1830s (Anderson, 1992, 85–109). Essentially, acting on data and adjusting a brew is a fundamental of good brewing practice. One could say that quality systems have been so ingrained in the science of brewing that it is hard to untie them. Still, some brewers continue to struggle to bring consistent beer to the customer, or evolve their approach to quality.

In the 1990s the craft brewing industry had an early peak and decline. Poor quality is sometimes the variable that is cited for this initial bubble burst (Trembley, 2005, 130). The question can be raised, "Why do some breweries struggle with managing quality and others thrive?" The answer may be those that thrive not only determine how to "do quality," but they understand *who* continuously leads the quality management efforts in the lifecycle of a brewery. This is a unique challenge in technical industries; transferring the technical and quality related decisions from one person in a start-up to the rest of the brewery staff must happen for quality to excel.

WHO – QUALITY FUNCTION IN BREWERIES

Creating a clear and focused role for quality in the brewery can be challenging, but it is essential! Quality management takes a special skill set of its own. For many small breweries the brewmaster typically wears many hats, and traditionally wore all the decision making ones. These duties may include the following areas: quality control, quality assurance, quality management systems, metrology (science of measurement), inspection, staff training, auditing, problem solving, reliability engineering, supplier quality (raw materials, including packaging), product formulation, and design (Westcott, 2005, 316–321). This is a large list, which may or may not be covered by a single person as a brewery expands. Like one circle that slowly becomes two, the splitting of roles can be gray. (See Figure 1.2.)

Bringing in new skills as a brewery grows is the best way to evolve quality. Many brewers have discovered that the formal "quality" skill set can be brought to brewing from other industries, such as automotive. Shouldering operations responsibilities, in addition to setting quality specifications and standards, the brewmaster must not only ensure the operations department meets the product demand, but also that the quality requirements are met. They may be called upon to troubleshoot when something goes wrong and also responsible for managing a staff safely. At some point, it becomes a lot to ask of one person.

Breweries still need to maintain process control and product quality as they grow, year after year. Adding capacity, new processes and technology, innovating with new products, and adding staff and complexity is a challenging job. Making sure the control systems and coordination aren't lost in the shuffle may be too much and too important to ask of one person. These challenges are not unique to small breweries growing today. Not too long ago, the megabreweries also went through major changes, consolidation, and growth, and at the same time changed the role of the brewing chemist to a quality manager.

Another key role in the area of quality is strategic planning. As risks change and evolve, a quality manager also wears the hat of a risk advisor for the business, especially if new products are a strategic area of growth. Most importantly, adding personnel to focus on quality will help establish a culture of consistently meeting the customer's expectations, a job the brewmaster traditionally held. In other words, hiring a quality manager as a brewery grows can not only help organize the control system, but also ensure it is coordinated.

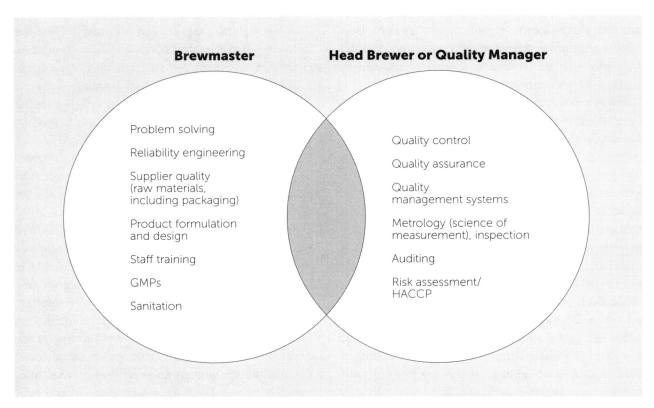

Figure 1.2: This figure highlights the overlapping job roles and responsibilities in brewing operations and quality management. If a brewery is very young, the circles may actually be positioned on top of one another. As a brewery grows the circles become more distinct, and job duties become carved out. This is just for illustration purposes. The reader's brewery may operate differently, but it is key to begin to consider these roles and how they will section off as the brewery begins to grow. This graphic illustrates how difficult it can be to carve out roles and responsibilities that ultimately overlap. Regardless of how key responsibilities split out, both quality manager and brewmaster have support roles for one another, thus they overlap. In other words, if supplier quality is a brewmaster responsibility, it is the quality manager's role to support the brewmaster, not usurp them.

In addition to the brewmaster, at some point during a brewery's growth it makes sense to bring in laboratory expertise. The brewing chemist, who also may function as a quality manager, may carry some of the same responsibilities as the brewmaster. Like the brewmaster, the chemist must have command of the process, the requirements of product quality, and process control. The chemist may also have a special knowledge of the control system and how measurement issues may arise, and has to partner with the operations teams to troubleshoot and be sure product quality or control was not lost in certain circumstances. In other words, the brewmaster and chemist must not only be aligned to the product and process standards, but also how to coordinate the control system and communicate results. This responsibility overlap can cause confusion. It can certainly be a point of pain if the chemist is usurped and defective beer is released to the customer before being properly evaluated and disposed of. Exploring the history of this relationship is helpful in making the case for evolving the role of the brewmaster as the brewery grows and expands.

The person in the position to organize the quality control system and oversee its coordination must also have the authority to make changes to the quality control system if necessary. That person should wear the quality manager hat, and should be considered a partner with the operations leaders in bringing forth the best quality product. If the brewmaster is filling this role, be sure the brewery isn't under the stress of adding things such as new process controls in the cellar, coordinating new tank arrival, or designing and managing all aspects of a new product. In other words, adding growth concerns to their plate will be too much to manage and the quality system will suffer and not evolve.

The Brewing Chemist

This history of brewing has its ups and downs. From prohibition to WWII there were clear interruptions to the brewing industry. In the 1940s and '50s quality management began to flourish as a separate discipline, and breweries were transforming their business as well. The role of the chemist was about to change. As mentioned previously, being a highly technical trade, breweries already had chemists and even statisticians in the operation. During this period of time, breweries were under a tremendous new wave of consolidation and building large facilities. It was noted that breweries were putting up significantly more capital compared with other industries.

The capital expenditures in 1947 were published to be $110 million—fivefold the amount they were in 1939. "The only other industries listed as spending comparable amounts for the same purpose in that year are sawmills and planing mills, paper and board mills, and certain key industries…" (Baron, 1962, 278, 339). The capital investment allowed their breweries to expand and grow, but it also led to adding staff, changing technology, and other new challenges (McCabe, 2005). *Bloomberg Businessweek* published an article in 1947 relating that as breweries grow and change, their methods also must grow with the times. This was re-published by the MBAA in 1999 ("Shifts in Beer Picture" of the July 17 issue [number 98] of *Bloomberg Businessweek*).

> "Needed: New methods—Regional and shipping brewers have some problems in common. One of the main ones is the old-fashioned approach to their business. Many companies still consider beer making an esoteric craft; they adhere religiously to the same inefficient methods used by nineteenth century German brewmasters. The larger breweries are gradually bringing in new production-line methods for turning out the suds. In many breweries the chemist is replacing the brewmaster as top production man. The brewmaster merely carries out the chemist's instructions." (McCabe, 1995, 241-243)

The *Newsweek* journalist was pointing out that as the industry matures, breweries need to update the old matrix of responsibilities. One can imagine the suggestion that "the brewmaster carries out the chemist's instructions" caused a stir. A comment on this article stated:

> *The implication is that a brewmaster is not a chemist. This is a ridiculous assumption simply because the skilled brewmaster is well versed in all of the chemistry as applied to brewing.… The capability of the Master Brewer to investigate, interpret, and apply the results obtained through all means and methods known for brewing control and development constitutes what is known as 'The Art of Brewing.'* (McCabe, 1995, 241-243)

This commenter's suggestion that the brewmaster must run the show in breweries, establishing standard processes, troubleshooting, and working closely with their right-hand person, the chemist/microbiologist, missed the message for a paradigm change. The *Newsweek* article's author postulated there is a need for "industrial skills beyond brewing." This is the key feature, and it was a growing tension then and now. The brewmaster may bring the technical knowledge and flavor sensitivity needed for crafting delicious beer, however, the industrial skills to monitor and correct, problem solve, and keep quality contained may not reside with the brewmaster. The previously mentioned 1953 article by Brenner originally also touched on this personnel conflict:

> We must work according to sound and reasonable principle, with sound and reasonable equipment, with sound and reasonable personnel. We must have a man responsible for general supervision and planning, normally the executive in charge of production…. The chemist should be free for special investigations, for seeking out the sore spots, for planning the details of the effort, and for doing involved and difficult laboratory work, while the technician carries out the work-a-day routine. (Brenner, 1996, 193-198)

Interestingly, this natural split in roles has also been played out in industry associations. The American Society of Brewing Chemists (ASBC) had a close tie in history with its partners at the Master Brewers Association of the Americas (MBAA). About the same time the principles of quality control statistics were being established in other industries in the late 1930s and '40s, the brewing chemists began their road to independence as a separate group with a clear mission in their association. Sanctioned as a committee in 1934 by the MBAA and the US Brewers Association[1], ASBC chemists initially began by standardizing malt analysis back in the early '30s. The ASBC became its own association in 1935, establishing bylaws and methods of analysis. It is notable how tied together ASBC and MBAA were even back when the comment from *Newsweek* came out, and how quality came right in the middle. In 1945, a liaison committee between the ASBC and the MBAA had to charter to expound upon "The Keeping Qualities of Beer." (Tenny, 1984)

The Chemist and the Quality Team
The separation of the associations allowed for a division of responsibilities within the brewery. The brewmaster no longer dictated the method of measurement; instead, the brewing chemists developed standard methods of analysis for the industry. The chemist's responsibility to the brewery is well anchored as one who ensures the measurement system is healthy. Still, as quality management grew into its own discipline with Juran and Crosby leading the charge, how the brewery utilized the senior chemist beyond the lab depended upon whether the brewery had the time to focus on evolving quality into a plant-wide responsibility. As other industries were building their organizations based on the total quality management mentality to reduce costs, and costly quality errors, breweries were under pressure to scale to greater heights.

The 1960s and '70s saw tremendous growth as the major breweries developed national footprints. The gurus told manufacturers to share the responsibility of quality in order to troubleshoot more effectively, lower costs from waste, and improve output. In other words, shrink the quality team and add the responsibility of measurement to everyone's plate. But at that time, quality teams at breweries were inflating. The Miller archives note a quality team expansion from 1974 through 1978 from 100 to 530 personnel. "Long time suppliers were surprised to find themselves receiving detailed analytical reports on the quality of their products, whether they were selling malted barley or can lids" (Gurda, 2006, 134, 156). This was a divergence from what Juran and some of the quality gurus were preaching. Digging further, the inflation of a quality department in breweries was likely a necessary evil to their growth.

Back in the 1960s and '70s, the large brewery consolidations and simultaneous growth combined operations to large networks for efficiency, but left little time to train a large operations staff on conducting quality checks as an additional part of their job. Keeping a quality staff on top of the checks made sense, but the regional and multi-plant breweries of those decades were mostly staffed with unionized labor, with which they had an adverse history. Occasional strikes and walkouts that did occur were problematic (Gurda, 2006, 134, 156). Transitioning quality check duties from laboratory technicians to brewing staff during periods of unrest was likely not advised. The labor issues that cropped up during those transitory decades may have factored in to whether or not the principles of total quality management were adopted by a brewery.

The Change to Quality Is Everyone's Responsibility
During the 1990s at the major breweries, large quality departments went out of vogue. Waste, inefficiency, and cost control became extremely important as the competition for the top heated up. The new direction for quality departments of the major breweries was to transition their inspection and measurement duties to the front lines and brewery staff, and become subject matter experts on the difficult measurement and troubleshooting duties. Many breweries recognized the need to execute consistency in product; quality had to become every person's responsibility, not just the brewmaster's or the chemist's, requiring a sharing of science discipline. There is much evidence in brewing literature of a new wave of adopting TQM and the recognition that waste was no longer acceptable.

---
[1] Not related to the current day Brewers Association.

The Schlitz Beer Quality Story

Schlitz fell from prominence during the early 1970s. Historians attribute the decline to many reasons, but quality is cited frequently. In the late '60s Schlitz introduced a process called Accelerated Batch Fermentation (ABF) to improve their brewery utilization. It was a much lauded program at Schlitz, as they thought it showed how efficient their brewery had become. However, Schlitz had also changed raw materials, substituting corn syrup for malted barley as one of many cost-cutting measures. In other words, Schlitz leadership chose to compete strategically on cost, which resulted in changes to the brewing, fermenting, and aging processes that needed to be managed closely. After the changes, complaints from distributors began coming in, citing inconsistent flavor and foam retention (Trembley, 2005, 130). Instead of fixing the root cause of the complaints (possibly a combination of accelerated batch fermentation and raw material substitutions), Schlitz decided to add a foaming aid, and replace a clarifying aid with silica gel. The combination of those two substances resulted in flakes in the beer (Skilnik, 2006, 246). This became a tailspin of quality failures. It started with a lack of identifying what was critical to the customer, built upon itself with a lack of clear process control stops, and ended with troubleshooting in a vacuum, resulting in significant downstream issues in finished goods.

The Schlitz story shows why adopting the quality gurus's advice on quality management can save a brand. One might wonder if Schlitz had firm policy on listening to the customer (defining quality as lack of defects), ensuring explicit quality management oversight, and incorporating brewery-wide problem solving with appropriate personnel at the time of their tailspin. Knowing the challenges breweries faced at that time, perhaps some elements were not as robust as they could have been.

From the MBAA archives, Hull wrote on TQM, "Long a traditional industry, brewing has entered into a period of intense competition and rapid change where the performance level required to remain competitive has become a moving target. Total Quality Management provides the new management framework needed in a market where success depends on continuous improvement" (Hull, 1990, 42–46). South African Breweries (SAB) was known for the management philosophy of World Class Manufacturing during the late 1980s and early '90s, a period of incredible change in South Africa (Carter, 2003, 121). At the other major breweries in North America, TQM became the buzz word (Goff, 1995, 24).

Growing breweries today that define quality beer in more of an esoteric manner should look to the inflection point from the 1980s to '90s in the industry and learn from the past. Certainly the brewery economics of the '80s and '90s imparted the need to change the model in large breweries and for the quality team and the brewery to become more efficient in the operation. The challenge for today's small breweries is to avoid sliding back in history and repeating the failure to establish a modern quality management practice right away. The challenge is that there may not be a compelling financial reason to evolve from the old model of the chemist being the only quality control entity, and the brewmaster calling all the shots.

The thought may be, "If all is well, don't change a thing." It costs money to swing the quality chemists into a role of teacher, troubleshooting expert, and strategic operations partner as modern quality management principles guide. If sales figures show tremendous growth, then meeting production needs is the first priority. However, brewery leaders need to look beyond the rate of growth as evidence that all is well in quality. As competition heats up again, a large, defective product loss will not only be very costly, but will also quickly shift a customer away to another brand, as happened with Schlitz. Leaders need to give equal weight to the cost of consumer complaints, added complexity and the requirement for a swelling quality department, and the risk of product recalls. The benefits are clear that creating a quality culture, one that evolves with the growth of the brewery and develops the chemist and brewmaster's quality management roles jointly and strategically, is vital.

However, brewery leaders need to look beyond the rate of growth as evidence that all is well in quality. As competition heats up again, a large, defective product loss will not only be very costly, but will also quickly shift a customer away to another brand, as happened with Schlitz. Leaders need to give equal weight to the cost of consumer complaints, added complexity and the requirement for a swelling quality department, and the risk of product recalls.

Once it is decided that quality culture is strategic, the natural tug-of-war between quality and operations can cease. Instead, broader synergies in quality processes can be implemented. The quality management role—perhaps the chemist, perhaps the brewmaster—becomes one of the partners in the brewery operations. They will add discipline to measurement and problem solving, communicate metrics, and maintain consistency of product for the customer. Their job is to be an advocate for the customer. Today's brewers may not have the many challenges with consolidation and labor issues that plagued the megabreweries during the heyday of their growth, but, as before, growth and innovation add new product quality risks into the brewery and they must be properly monitored. If the lessons of the past tell us anything, it's that it is easy to get distracted from the evolution of quality culture during growth.

The brewing industry is not all that different from other industries in that quality is: standardization of process, clear and defined specifications based on customer needs, process control with measurement and standard corrective actions, continuous improvement, and empowerment of control at the most fundamental operation level. If a brewery takes the time to define its quality values and customer criteria, and gives employees responsibility for quality, it will allow a quality control system to evolve with the changing dynamic of the business. There is a large upside to all of this, so let's get started.

TWO

QUALITY MANAGEMENT AND GOVERNANCE

Some breweries have a well-established quality management program in place. This is clear from how these breweries run; quality staff and operations often working exceptionally well together. Any quality improvements happen jointly with new capabilities and new products, and, most importantly, accountability for the quality of the product is felt at every level. This sounds like an ideal state to some. Unfortunately, as mentioned in the previous chapter, there can be issues in a brewery between the quality staff and operations staff, the splitting of duties and authority. This can happen in a brewery that is experiencing change or managing growth, as well as in breweries that are not growing but that haven't completed the fundamental steps of establishing quality governance. This chapter will explain the difference between quality governance and quality management and illustrate that by making a clear distinction between the two, the brewery achieves a well-rounded quality culture of logic and control, versus confusion and chaos.

Connecting the Dots Between Management and Governance

Two key responsibilities require management oversight: the responsibility to define and set the boundaries of a product's quality criteria and specifications, and the responsibility to execute the quality system and improve on the brewing, packaging, and delivery process. Quality governance is the setting of policy, strategy, specifications, and goals for the brewery as they pertain to the quality of the products. Quality management is executing this policy via the measurement and control plan (or the "quality system"), resulting in process and product control, and great tasting beer.

At the connection of quality governance and quality management there must be communication, training, and direction in order to determine where to put efforts toward improvements. This "in-between" layer can swing to either the governance team (who may be an upper-level management team dedicated to operations), or the operations team (the brewery management). The activities must happen regardless of the size of the organization. Not defining who maintains the responsibility to communicate, train, and direct is a grave mistake. As operations can change over time, the governance team can no longer manage all the analysis, communications, training, and improvements. This is when decisions must be made regarding the roles and responsibilities of outlining a quality system.

Let's dig further into quality governance and quality management so the reader can make a more informed decision on how to manage these roles. We will explain how to plan for each in a brewing environment. Lastly, we will also walk through the process of writing a quality manual for a brewery, so all governance and management duties are explicit to even the casual observer.

Quality governance is the setting of policy, strategy, specifications, and goals for the brewery as they pertain to the quality of the products. Quality management is executing this policy via the measurement and control plan (or the "quality system"), resulting in process and product control, and great tasting beer.

QUALITY GOVERNANCE – STATING WHAT IS EXPECTED

Quality governance is "setting the expectation." In small breweries it can be unclear who is determining the rules and who is simply following them. The rule setters must have the undeniable authority to set rules, specifications, or expectations of what the operations have to achieve. Laying out the rules is the responsibility of the person or team that governs.

Those responsible for quality governance answer questions such as:

- What is the expected outcome of the brewery in terms of quality performance? (Total customer complaints? Lost beer to quality issues? Beer pulled from the market due to spoilage?)
- What does "good" look like in our behavior and culture regarding our product's quality? (Do we say what we do and do what we say?)
- Who has what decision-making authority toward quality issues? (Front-line brewers, quality manager, brewmaster, etc.)
- What is the policy (or philosophy) of the brewery leadership about quality? (Do they encourage making the right decision even if it is a hard decision? How do they encourage this behavior?)

In both large and small organizations, a corporate quality group or some form of centralized governing body sets the rules. If a company has more than one site, leadership must decide whether centralization or decentralization of quality governance is best for the organization. It would be highly unusual, but not unheard of, to have several independent quality managers govern their breweries, or perform rule setting on quality policy. This may occur if the brewery conglomerate has divergent sizes of breweries in their network, or very different types of beer being made in their breweries. If this is the case, decentralized governance may be the more efficient means to set rules and govern.

In a small organization, the governing body for quality may be simply the brewmaster and the head laboratory chemist setting up the measurement system. This would be an example of centralized governance. This responsibility can be given to a one- or two-person team versus a large team. Governance or rule setting will likely remain centralized to an individual, or a small team, as long as the brewery remains a single entity and the processes are similar enough. The size of the team isn't an issue. Trouble begins when the governance responsibility changes hands, as a brewery grows and adds complexity. For example, rule setting may be transitioned from a brewmaster or brewery owner to a quality department head. Without proper sign-offs and agreements for delegating decision making, confusion may result.

The other area of trouble in smaller organizations is when the person or team setting quality standards also has operations management duties. In these cases, clear decision rights of changing a company policy or changing a quality specification must be stated. Additionally, escalation for disagreements must also have a clear process. (More on delegation and escalation later.) As you can imagine, there is potential for a conflict of interest when an individual is managing a production schedule and brewing operations, and is also responsible for establishing the quality policy, specifications, and procedures. It is simply too easy to trounce over that policy when convenient. If your organization is too small to split up these duties, then the owner or founder must be explicit in how changes in policy and specifications are to be managed. Setting policy takes willpower and tenacity; if conflicts of interest are not well managed, all that work can go down the drain.

The Quality Manual – Setting Policy, Specifications, and Goals

Policy, whether it is human resources, financial, or quality-related, usually starts with senior leadership, whether or not they know it. By expressing expectations verbally or in written form, a leader is instituting a form of a policy. The brewmaster or operations leaders may translate the leader's expectations to the rest of the brewery through actions. As the organization grows, there comes a point at which company policy should be formalized, and written. Specifically, quality policy should be formalized as soon as there are more than three people that have to interpret what the quality values and goals are. If you have ever played a game of telephone, you know why. After two or three interpretations, the original message can get lost.

Quality policy is typically part of the quality manual. The manual will have a policy or value statement, and also explain the organizational structure and goals toward quality. A quality manual clarifies to the reader

Quality Governance

The American Society of Quality's *Global State of Quality* published in 2013 organized "quality governance" in five distinct groups. This illustrates the complexity of governance across many industries.

- Centralized (one quality department oversees a single or multiple plants)
- Localized leadership at multiple businesses (quality leaders at each plant)
- Centralized committee made up of multiple functions (such as the brewmaster, head brewer, packaging manager, and lab technician)
- Senior leadership governance (the owner/president runs quality)
- Corporate organizational board of external representatives (such as an adviser in the brewery)

Survey respondents included multinational, very large organizations and smaller single-site organizations in over 16 countries, so it is a broad view of the state of quality. In 2013, 35% of manufacturing respondents to this survey selected a "centralized quality department" as their form of governance. A decentralized authority for quality governance is not common in manufacturing, especially a technical manufacturing operation that has a similar process in multiple locations. Decentralized quality governance is more likely found in large organizations in which there are several different technologies or there are many small companies under one parent company. A centralized quality department as the governing quality authority can be found in large multinational breweries today. However, some sites choose a different path. For example, a committee of leaders from multiple departments (such as packaging, brewing, and human resources), or multiple sites (such as many quality managers across multiple breweries) function as the quality governing body. In other words, there isn't one size that fits all in governance. (ASQ, 2013) how the brewery will manage the quality system, and who has governance over the specifications. It may include goals regarding consumer complaints or consumer satisfaction. Lastly, a manual clarifies how the escalation of quality-related issues is to be managed. If the brewery is seeking some sort of accreditation for its quality system, this manual will become an extremely important document and should reflect the specific accreditation requirements. It is essentially the evidence for "say what you do."

> **Writing a Quality Manual Follows a Simple Process**
>
> 1. Articulate the company's values in terms of quality for the product and organization in a policy statement.
>
> 2. State the roles and responsibilities toward quality governance and management (who is accountable for what, and what the vision is for future responsibilities).
>
> 3. State the specifications for finished products and the frequency of review.
>
> 4. Write out all procedures and actions for when a specification is breached. (These procedures may take a while to establish—more on this later.)
>
> For an example of a quality manual template, see Appendix A.

Since "policy" is an all-encompassing word, it is easy to use it broadly. For our purposes here, quality policy is the way to behave, and to honor and execute the quality system. To write a policy, many small companies bring in an outside contractor to help get things started. Setting policy isn't something those who start a new brewery necessarily get excited about. It is administrative work that can have great impact. It involves writing down the company's values and vision for quality. When done well, it becomes an operation's guiding light. Who physically writes the policy isn't as important as who *leads* the policy decisions. This must be the most possible senior leader that is knowledgeable about operations.

Articulating a company's quality policy in one or two sentences may seem difficult at first, and for some seems like an exercise in futility. However, the need for quality policy becomes apparent when trade-offs emerge, leadership changes, and growth happens. The values of the company, usually stated elsewhere, are still lived through policy, even the quality policy. Therefore, the policy statement is not just for show, but a guidance tool for management and employees as an extension of already inherent values. For example, if a company values sustainability, then the policy may say, "Our quality policy is to balance the delivery of world-class beer while ensuring sustainable practices are utilized where feasible in brewing, packaging, and shipping operations." The brewery that states this as policy will not be at odds when determining how to improve quality if, for example, quality and sustainability require a trade-off. Let's say, for illustration, a brewery has an initiative to save energy and is considering cutting down the heating time of a chemical in the cellar's clean-in-place (CIP) process. This initiative may impact product quality. Therefore, as part of policy, the initiative would require quality testing and have performance metrics related to quality and energy to assist in the decision-making process. The result is a respectful balance of maintaining quality while continuously striving to find new ways to lower energy costs.

> **Example of Quality Policy for Sustainability**
>
> "Our company values sustainable practices. Therefore, our quality policy is to balance the delivery of world-class beer while ensuring sustainable practices are utilized where feasible in brewing, packaging, and shipping operations."

The policy statement may evolve and morph with the brewery's maturity and growth, so the best advice is to make the statement complementary and aligned to the rest of the organization's values at the point it is written. Company values that are in alignment to a quality policy are:

- Learning environment
- Innovative environment
- Continuous improvement
- Empowered employees
- Structured auditing
- Respect for community and customer, or local initiatives
- Sustainability

Note that this list does not say things such as "We value sourcing the cheapest ingredients that may impact quality," or "We value efficiency at the price of quality-tasting beer." Making sure the value of the product is inherent in the quality should ring true in a policy statement and should feel comfortable to live with, no matter the trade-offs. Taking the time to articulate the quality policy will help balance the trade-offs and be the guiding point for the difficult decisions that most operations face at some point.

An example of a policy statement from a company that values employee input would be: "Our quality policy can be summed up in the value of our people. Our brewery believes our people make quality decisions, and they value providing the best tasting beer to our customers."

The Quality Manual – Roles and Responsibilities
We have already mentioned quite a bit about roles and responsibilities for quality governance in this chapter.

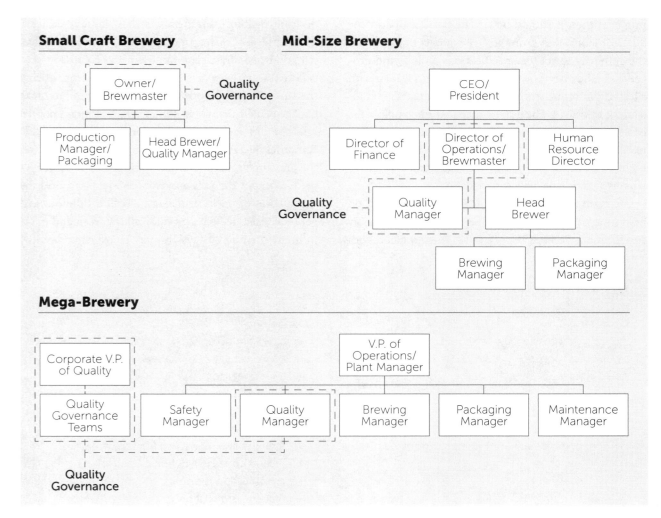

Figure 2.1: Examples of various types of organizational charts depicting quality governance and management.

However, note that the manual should clarify who does what in the realm of all quality tasks in governing and managing (see sidebar). Typically, an organizational chart for the quality team or managing team is represented in the manual as well. This isn't required, but is a nice, helpful image for those who are new to the organization. The manual is a way for new employees, auditors, or others to see what the team looks at through the "quality lens." An example of typical organization structures can be found in Figure 2.1.

The roles and responsibilities section should also address delegation and escalation. As the original brewery owners become more involved with the broader business, or the brewmaster's duties grow, they may delegate the total responsibility for governance to the quality department. If the delegation happened properly, then it should be updated and stated in the quality manual. For example, "The quality manager has the authority to set the specifications, policy, and procedures. They perform this duty as a team leader, with support or input from the operations teams, senior leaders, or others. The quality manager must follow the escalation process if a specification is questioned." The brewmaster, either as the leader of the quality department or a peer to the quality manager, must adhere to what is stated, especially for escalation.

For example, let's assume a brewery has product on hold in a finished beer tank due to color that is below specification. Perhaps they have no beer to blend with the product on hold, and have to wait two weeks before the next beer is ready to blend in color. The quality team discovered the specification breach and, following protocol, put the tank on hold. They informed brewing operations and scheduling. If the brewmaster helped set the specification, as part of the governing team their best response is to support the decision and work on solving the root cause of what happened to the off-color batch. If the governance wasn't properly granted to the quality manager, the brewmaster can possibly ignore the decision made by the quality manager and release suspect product. This product is not a safety issue for the consumer, but the release may result in consumer complaints. Holding, blending, and focusing on the root cause of the error is the best course of action.

In this case, if there were a disagreement between decision-making team members, they should escalate to a higher level of management, even the owner if needed. All good intentions crumble when the escalation procedure doesn't exist or isn't followed. The danger being that any person with power will be called upon to make quality-related decisions, sometimes usurping another. Like the old saying goes, "If Mom says no, ask Dad." Breweries, just like any manufacturing business, can feel the pressure to fill orders, and if the gap exists, rules get bent. Without the escalation process in place, and the expectation of decision makers to call a time-out and get final clearance on a product, all the work and effort in the written quality system may be overshadowed by

---

Roles and Responsibilities

Governing
Typically conducted by senior leaders, technical leaders, or corporate leaders (in large brewery conglomerates)

- Setting specifications
- Establishing measurement procedures
- Changing policy or procedures, governance delegation
- Escalation
- Goal-setting and quality improvements, including consumer complaint reduction

Managing
Typically conducted by those managing operations such as production managers, quality managers, and line supervisors

- Measuring and reporting (quality control and quality assurance)
- Training and managing human resources
- Taking corrective action on out-of-specification products
- Data review and continuous improvement from all data, including consumer complaints

the push and pull of personalities. If that is the case, the customer, and eventually the business, will feel it.

The escalation process is a way to resolve natural conflict or disagreement when it comes to interpreting policy, or specifications. Usually, there is room for interpretation in written rules. Therefore, it is best to prepare for this inevitability with a simple process. An example of an escalation process is, "If a product hold or disposal or a policy interpretation is questioned by others, escalation is brought to (insert your appropriate senior leader's title here) for a timely decision. If policy must be changed because of new input, senior leaders must sign off on the recommended alterations."

The escalation process requires any manager to have the maturity and responsibility to say, "We disagree, so let's take this up a level and get it resolved." Therefore, as governing duties get transitioned, the operation needs to make it very clear how disagreements are escalated and who has what authority, in the case of a specification breach, to determine disposal or a change in usage for a product. Providing a trump card to one person is not recommended. Instead, a clear escalation process should be agreed upon with the company owners or leaders. Any decision to breach written specifications must be seen as a very serious decision, one that must be signed off on by the highest authority. In a large company this may be a chief operating officer (COO) or a vice president (VP) of quality. In a small company this may be the president.

This seems like a lot of red tape and words, especially for a small brewery that has a lot to juggle. However, good working relationships are challenged when beer is on hold. If product is at risk, work is questioned, and pride is at stake. It is emotional. Take the emotion out of the picture by having a well-written procedure to review the presenting information against current specifications and inform parties of discrepancies without a lot of drama. If disagreements happen, then escalation is just part of doing business.

When a quality manual is written, the roles and responsibilities segment usually relates to the current point in time. At some point in the future, however, the management of the quality system may look very different. It is a good idea to establish some sense of what the future will look like so there isn't conflict once the organization starts moving in that direction. For example, "It is our company's primary quality strategy to continuously improve our operation. We envision our brewery will have self-managed and well-trained teams that effectively manage the quality of their process and outputs." This statement supports the teams that are working toward this vision. If a cellar team is self-sustaining they train each other, they continuously improve their processes, their quality output is excellent, and, if they communicate well, no one should be threatened by their activity. That team reached the ultimate goal of the brewery, to be self-sufficient in quality. They should be commended.

The last part of a quality manual should have written specifications and standard procedures. We will delve into this subject later in the book. For now, simply note that though this part of the manual is important, and while it can be overwhelming to get started, it will evolve with time. Therefore, it is important simply to get started; get the basic product and process specifications written down and be prepared to add to it as part of routine quality system maintenance. Assuming the right homework was done prior to this step, you should know who owns the specification review process as part of governing quality. It is important to have the *who* in place before the *what* in this circumstance. The quality manual really takes foothold after specifications and procedures are written. Placing the manual in the right hands makes it come alive.

QUALITY MANAGEMENT – IMPLEMENTATION OF POLICY

Quality management is the implementation of the policy, specifications, and goals set by the governing body. It usually requires measuring and assuring the quality of the data. In addition, a quality management duty is to create the strategy to improve the operations and measurement processes in tandem. There may also be regulatory compliance and training responsibilities embedded in quality management. In most breweries, the quality management job responsibility is considered part of the operational responsibilities. Summarizing, the quality management responsibilities are:

- Measuring and reporting (quality control and quality assurance)
- Training

- Taking corrective action
- Strategy to review and improve or continuous improvement
- Regulatory compliance management

How the organization is designed to execute quality management depends on the brewery, its level of complexity, and its philosophy toward quality. The brewery may or may not have a dedicated quality manager. Individual titles will not matter if every manager is considered a quality manager to some degree. The quality management roles are drilled into the organization by assuming every manager or employee has the responsibility to measure the process they run, ensure their process and product meet the specifications, and improve their process if specifications are at risk.

Quality Control and Quality Assurance
Let's first establish the difference between two arguably close-sounding terms in quality management: quality control (QC) and quality assurance (QA). Both QC and QA are executional duties, or in other words, there is a routine duty to execute. QC is controlling the process and product output by conducting a measurement, usually as close to the operation as possible. QA is controlling the measurement output by assuring the data via calibration or checking a *measurement* tool with a standard. QC duties can fall directly on the operations staff running the process, or they may fall on a quality lab technician or supervisor. QA and QC processes intermingle. For example, measuring pH in wort is a common practice during the brewhouse operation. The brewhouse staff is typically responsible for measuring the pH and adjusting the mash liquor (QC). They are trained, by either a quality technician or a peer, on how to measure pH, and they also learn what to do if a measurement does not fall within the specifications (QA). They will likely first re-test the sample, calibrate their measurement equipment, and then adjust the process if needed (QC). If the pH probe needs to be serviced, or the measurement accuracy and precision need to be validated, the quality technician may help (QA).

A brewery usually has fewer QA resources than QC resources. Note: If calibration is required routinely for instrumentation to work properly, it should be part of QC duties. However, monitoring calibration results, servicing instrumentation, and validating measurements on a broader scale is QA. The QA check for pH may be monitoring the calibration results for the instruments on the floor, and spot-checking production staff processes. QA, in other words, is part of the safety net to ensure quality checks remain in control. Either an operations team or a quality team might manage QA. Again, who does the work is up to the brewery management to decide. Regardless, the QA duties must be completed for the quality system to be robust.

> **Measurement Systems Analysis and Quality Assurance**
>
> Gage Reliability and Repeatability (Gage R&R) is a type of statistical analysis that provides a level of accuracy and precision of a measurement system. Some breweries elevate their QA program to include this level of sophistication to gauge the measurement system as part of a Six Sigma initiative. It is a recommended process if the brewery is on the continuous improvement path to reduce process variation. Gage R&R is a helpful means to ensure that measurement is not the source of variation in a process.

Training
Training is another management duty that is required for a quality system to work really well. QC training responsibilities may fall on the staff members (to train each other) or on a supervisor (to train a staff member). The QC training should be validated before staff members are allowed to run QC checks on their own. Usually the quality department has this role, but not always. It can be considered a QA duty to validate training.

Lastly, to be consistent in policy, specifications, and actions, it is good practice to "train the trainer." Take the time to train supervisors and other staff who routinely conduct process training on the quality policy and principles. Train-the-trainer practice should be required of an all-encompassing brewery training plan when growth extends to include multiple shifts and staff members. It should also be part of an all-level management orientation process, no matter what size the brewery.

Corrective Action

Another quality management duty is taking corrective action. This means stopping a process if a quality parameter exceeds a specification, correcting it, and then documenting the action. This should be part of QC duties and not left to a separate quality department. Any member of the operations team should be able to take corrective action on their process. It is inherent in quality system best practices that corrective action be taken as quickly as possible. In other words, the staff member is fully empowered to measure the process, stop the process if it is out of specification, adjust it, or take corrective action. This is called "quality at the source" and it effectively reduces the time to detect and correct a product quality issue. The less time that lapses between error detection and correction, the better the quality of the beer will be.

Take, for example, measuring CO_2 levels in beer on the packaging line. If samples are taken and set aside, and CO_2 checks are conducted after the operation is completed (say the next day), any specification breach in CO_2 levels from a poor valve or other root cause may be detected after the product is already packaged and shipped. This becomes costly to retrieve, or worse, it results in loss of customers if left in the market. Building a QC program to detect and correct issues as soon as a deviation from specifications is noticed is ideal, but it requires planning, training, and agreement across the board with operations staff. Training on the standard corrective action should be part of the training duties, and the documenting of the corrective actions should also be standardized. Most importantly, taking the corrective action must be considered a job well done, and not penalized. Corrective actions, and quality at the source, if well managed, are vital components to maintaining quality—and keeping costs low.

Some "quality at the source" tests in a brewery can be difficult to justify. Breweries are unique environments that sometimes have very manual measurements throughout the process to maintain control. Usually the easiest tests, such as pH, Plato, and even taste and smell are simple to train staff for and make part of their QC duty. But what about manually counting yeast cells to determine how much yeast to pitch in a tank? Or running a gas chromatograph to determine the diacetyl level and make a judgment call if a tank is ready after fermentation? The more complicated the test, the more likelihood for error. Making decisions on bad data can be a disaster for product quality. Some breweries are choosing to keep the more complicated QC tests in the hands of a chemist or microbiologist. Others are pressing forward and picking off the complicated tests one at a time and transitioning them to operations. It comes down to cost, the size of the quality team, and the skills inherent in the operations staff. It is a decision that has to be considered and made by the managing leaders. This ultimately means it is part of a routine discussion between operations and quality staff on what tests will be moved from the chemist to operations.

Corrective actions, and quality at the source, if well managed, are vital components to maintaining quality—and keeping costs low.

When the decision is made to transition a quality check test to operations, the training, in the form of operator-to-operator and written instructions, must also include the entire body of knowledge that goes along with the test protocol, including what to do if there is a specification breach. An example would be if pH checks are frequently performed by one brewer, but to save time, the packaging manager requests to conduct a check of the pH of the bright tank so he can get packaging a tank faster. The initial training should include the background information of how to clean and maintain the pH probe, and how to determine if it is not reading correctly by checking the standards. The training should include what exactly he needs to do, and check again, if the pH is too low or high for example. In this case, the packaging manager has now taken over the check and is given the authority to handle re-scheduling, troubleshooting, and other production issues if the pH specification is breached.

Any staff member conducting quality checks must know the "why" behind the corrective actions they will perform if they are given the responsibility. Assessing staff for gaps in knowledge of the quality check process, and the corrections they must take, is an excellent duty for the head of quality. The quality supervisor should also provide the plan when quality checks should be transferred between the lab or the operations teams. As less quality control time is left to a lab staff, they can transition into quality assurance, or facilitate continuous improvement and problem solving, therefore benefiting the brewery even further. To commit to "quality at the source" is to commit to employee empowerment. There are challenges in an employee-empowered culture (see sidebar), but it is well worth it.

Continuous Improvement

Continuous improvement (CI) is also a quality management duty. It can be lumped in as part of QC or QA; however, it is a best practice as a separate discipline that is *applied* to QC and QA. Continuous improvement can be as simple as troubleshooting on the spot, correcting a process and changing it for the next batch. Or, it can be an arduous process of investigating a long-term issue such as excessive variation in pH control. In other words, short-term and long-term fixes are both part of continuous improvement. CI becomes more difficult to administrate as a brewery becomes larger. There are more problems to solve and more costs in chasing the wrong problems. In that case, it becomes a separate duty to coalesce, organize, and manage the CI process. This is part of quality management. Short-term fixes happen all the time

---

Empowering Employees and Furthering Quality Culture with "Quality at the Source"

It is helpful to know the hurdles to empowerment and a "quality at the source" culture. First, management must give up some of the power it traditionally held. Before embarking on changes in duties or decision rights, ensure there is alignment from all involved to make the behavioral changes that will be necessary. The likelihood to fail is high, especially in a technical field where management traditionally held a lot of power (e.g., brewing). If a brewmaster keeps playing a trump card to override written policy when it is convenient, no one, especially the person who ran the test or made the original call, will be empowered. If a senior leader wants to avoid safety rules and doesn't wear their hearing protection where it is required in the brewery, the manager of that area is not empowered to execute policy. If one element in the chain doesn't go along with the change to an empowered workforce, the whole house of cards falls down.

Next, empowering employees requires a plan to transfer authority with clear expectations (and consequences) of taking over the responsibility. If a production line staff member is fully trained and authorized to stop the line, yet doesn't when a quality check fails, there should be a disciplinary plan in place. Operations should have the responsibility to discipline accordingly.

Lastly, and most importantly, leaders need to be willing to change. They need to work for an empowered workforce. The leader's role becomes one of a supporter, not a dictator. This doesn't mean leaders don't make decisions; they still need to make key strategic decisions and administrate the workforce. Management must ask employees to invest in the process, and then the employees must be allowed to run and improve it. In a growing brewery, one in which the transfer of knowledge and decision rights must happen quickly and constantly, it is important to consider who has what authority in decision making with routine frequency, and plan transitions where necessary. This may mean having discussions with all stakeholders, making training plans, instituting laboratory training (including wider-scoped process training) and putting action plans together.

in brewing operations, which is favorable to maintain quality control. However, the trending of many short-term fixes may indicate a larger problem that must be addressed. Usually a person or a team has the accountability to review data of the daily fixes that are occurring, and pass judgment on what is important to address as a long-term fix for the sake of cost and efficiency or risk to product quality.

> Short-term fixes happen all the time in brewing operations, which is favorable to maintain quality control. However, the trending of many short-term fixes may indicate a larger problem that must be addressed.

Keep in mind that the skill to investigate a problem and get to a root cause is a real discipline of study. It takes years to master it. Six Sigma is a program that helps organize this skill, and builds the statistical knowledge and foundations as part of the training. Once a couple of people have mastered the skill of root cause investigation, the process can be shared. Usually a CI program is developed and standardized so the rest of the brewery can share information and improve. As part of a quality system, everyone should be trained on how to investigate root causes, document them, and communicate the fix. CI programs require commitment and real stamina in management because they take rigorous and continuous training, planning, communication, and coordination to see the process through. It is easy to set it aside for other initiatives that seem more important, such as new engineering projects that involve fewer people. However, if a brewery is growing rapidly, or is at a level where it reaches a large enough base of customers, organizing communications, data sharing, and standardizing improvement is required for product quality and cost efficiency.

It takes years to build a culture and establish a disciplined QA, QC, and continuous improvement program. These three components must be reviewed regularly in order to determine where the gaps exist. Stagnation of quality management is a real and potentially detrimental issue when it comes to evolving best practices in the brewery. True world-class brewery quality results when the entire organization agrees to develop the quality program as the brewery grows, and hires accordingly. QC has to be organized, standardized, and transitioned to operations. QA then has to be organized, standardized, and transitioned to operations. Lastly, CI has to be organized, standardized, and transitioned to operations. There comes a time when a quality manager, with the skills and expertise in data mining, continuous improvement, and risk management, can benefit a brewery tremendously as they can help determine where on the spectrum of maturity the brewery lies in QC, QA, and CI. The brewery quality manager can also bring many tactical and strategic elements to bridge the gap between policy, specifications, and execution. They should be considered a strategic partner in evolving the quality sophistication of the brewery. By default, they become a key person of change. It may be strange to think a quality manager's key role is to help in a transition of all the duties they help organize and standardize.

Quality Manager Body of Knowledge

The body of knowledge (BOK) or quality skills for a quality manager (QM), according to the American Society of Quality (ASQ), range from leadership to understanding the entire supply chain. There are many options for those interested in building this level of knowledge, such as studying to become a certified quality manager with the ASQ. Other options include becoming Six Sigma certified, or International Organization for Standardization (ISO) certified. Hiring a quality manager with a broad spectrum of management knowledge (and maybe even ASQ certification) is important for a growing brewery. Brewing knowledge and comprehension of quality specs alone are not enough to evolve a quality program after a certain period of growth.

Source: *American Society of Quality Website. http://asq.org/cert/manager-of-quality/bok*

In world-class companies, however, the quality team evolves to a training function only.

No matter what size a brewery is, making informed decisions on the governance and management about functional duties related to the quality program will help create organization and vision, and keep the peace in the organization. It will also keep the brewery team focused on the customer and personalize their impact on product quality. A quality manual is a nice way to sum up the roles and responsibilities to both governance and management. It is well worth the time spent.

KEY TAKEAWAYS

1000–15,000 BBL Brewery	15,000–150,000 BBL Brewery	150,000+ BBL Brewery
• Governance and management will overlap at this stage. Be aware of the role overlap. • Get started on the quality manual and plan to add to it routinely.	• Governance and management is starting to split. Take the time to define the distinct roles. • A broader vision of brewery organization is taking place. The quality vision is becoming important. • A quality leader role and skills need to be developed.	• Governance and management should have well-defined duties and obligations to one another. • QC, QA, and CI need to become definable, separately managed objectives.

THREE

THE COMPONENTS OF A QUALITY PROGRAM

After the brewery has established the basics in its quality manual, and governance and management duties are fairly well spelled out, they can now add the other key components: Good Manufacturing Processes (GMPs), Risk Assessment including HACCP, Quality Control Plan with Specifications, Metrics and Measures, and Quality Assurance Plan and Corrective Actions. These components are what many see as the heart of the quality program, and what those overseeing the quality program must be well versed in. Therefore, as part of this chapter, we will address the most important aspect of a quality program—the people executing it. We'll also be addressing the question, "What is the right amount of resources and skills for a brewery to support the quality program?" Additionally, we will discuss the specific daily duties or work of quality: conducting an overall process review of risks and controls (Risk Assessment); setting up metrics, specifications, and measures (Quality Control Plan); assuring test results (Quality Assurance); and standardizing reaction to out-of-specification conditions (Corrective Action). One effective way to visualize and communicate these and other quality program components is to create a visual hierarchy of components, one example of which is shown in Figure 3.1.

IT'S ALL ABOUT THE PEOPLE

When it comes to the quality staff in a brewery, the ideal can be summed up as having the right people who are trained, and in the right place at the right time. It can be easy to get tripped up on any one of those areas. From the smallest of start-ups to the largest conglomerate breweries and from those breweries losing market share to those experiencing growth, there is someone who is taking on the role

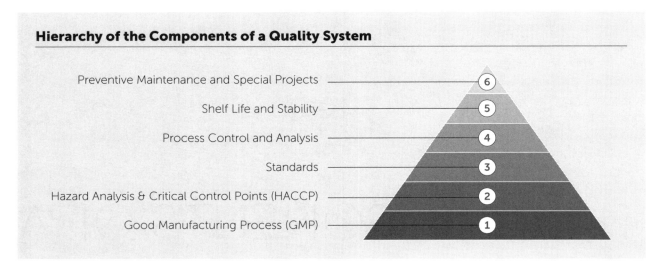

Figure 3.1: The Quality Priority Pyramid, created by the Brewers Association's Quality Subcommittee, illustrates how breweries may view the foundational elements of steps in developing a quality program. It provides a visual representation of the foundations of any quality program and which should be tackled first. Items farther up the pyramid are areas to optimize only after the foundations have been established. Reprinted by permission of the Brewers Association.

of product quality. Employees usually drawn to the job feel a strong inherent value to speak for the customer. They may have been on the other side of a bad customer experience, or feel very connected to the products. Those folks are the ones to tap for quality-specific duties when the brewery is large enough to have its own quality department. Those tapped for quality leadership not only have inherent quality values, but they must also have skills in brewing science and managing people through change. Quality leaders in the brewery will have to manage not only their department, but also the rest of the brewery staff through the development of quality as a culture. To help the reader understand what that entails, let's address the quality skills of the lab technician, and quality leader. How your brewery divvies up these skills among your staff is up to you. You may have brewery staff also perform the role of lab technician, for example. However, knowing what skills are expected for the role, and what gaps you have to address, can give you a great path for mapping staff development.

The Front Line – Brewery Staff
We will start at the front line, where quality has to happen anyway. This basic philosophy—that quality happens at the source of production—can be a hard reality for those used to taking every sample and conducting every check. It means giving up control, and it requires empowerment. (See Chapter Two.) The basic principle that quality happens at the front line, with the staff running the process, has to be firmly ingrained in everyone's mind. If a leader has any issue with this, they need to be reminded that performing the checks at the right location at the right time results in a shorter time to detect and correct an error in the process. A shorter time to correct an error means minimizing manufacturing losses due to quality failures. But there is also a side benefit, and that is the pride the staff carries when they are doing a whole job, not just running a machine. They are integrating themselves into the customer experience by conducting quality tests. The staff that conducts quality checks at their process also inherently know more about their process. This benefits the brewery because instead of one or two people understanding where the process may need some help, such as where it fails or where capital or investment is needed, there are many people. All of this breeds a community that contributes to improve quality in the product, and ultimately the brewery operation. This principle is called "quality at the source," and it has to be a value for the brewery to grow, expand, and improve.

We used an example of quality at the source in Chapter Two when discussing testing CO_2 in bottles or cans. There are many examples in breweries both large and small that implement quality at the source. The criteria to determine if a quality check is a good candidate for quality at the source are:

- Is the test easy to conduct?
- Does the test have a quick turnaround?
- Is there minimal technical knowledge and background needed?
- Can the test be run in a production environment or does it need to be run in a lab? (In other words, is the test robust enough, or is the equipment too sensitive to be on the production floor.)

If the answers to the above questions are yes, then proceed to move the test to the front line into operations. Checks that are commonly conducted at the source are pH, degrees Plato or gravity, CO_2, and air or dissolved oxygen. (Appendix B has a list of checks that are common to find in the brewery, and many can be used as a quality at the source check.)

To illustrate the thought process of considering if a check should be a quality at the source check, we will look at one check that can be a struggle for smaller operations. Assuming there are no in-line meters, the quality check for the yeast pitch process requires looking at yeast health under the microscope, counting the yeast, and sometimes sampling the yeast to plate for microbiology checks. It can be unclear who is best to perform the check. This is an easy test to conduct, and many times it has a quick turnaround. However, it may require more technical background training than the brewery is willing to do at the time. This is a critical check because if the yeast pitch quantity or quality is not up to standard, the fermentation will not proceed as expected and you'll have larger issues to deal with. It is not uncommon to see a lab person or a quality technician become deeply involved in yeast quality checking and become considered the resident subject expert. The yeast, if stored in a brink or from another fermentation tank, may be checked for microbiology (foreign yeast, bacteria), viability, vitality, and density of cells. Every yeast cell in the brewery has a story—its provenance, so to speak. What tank it was previously pitched into, how the previous fermentations performed, and how many times the yeast has been re-pitched are part of the yeast story.

Keeping track of the yeast pitch data can be tedious, but it needs to be properly documented. In addition, it is a special skill to look at yeast under the microscope, prepare dilutions to make counts, and interpret the data from prior fermentations. So, there are inherent risks of allowing this check to be conducted by too many individuals. Yet, if one person is a bottleneck in the process (e.g., they are on first shift and the third shift needs to pitch yeast), then the management has a decision to make: hire a third-shift quality person, or, instead, train the third shift and empower them to make decisions about yeast quality and quantity themselves. Both scenarios have a cost to the operation. The cellar staff member who now has to take 15–30 minutes to take a sample, count the yeast, calculate a pitch rate, etc., will be taking time away from their other shift duties. And there is obviously a cost to having a quality technician hired for a third-shift duty if there isn't other work they can do. This is the sort of decision brewery management has to weigh on a daily, weekly, or monthly basis, and there isn't one answer that works for every brewery. Usually, if the test resides with the quality chemists or microbiologists, it can be standardized and transitioned to operations at a later date. One hybrid solution is to have a quality point person on every shift that has both operations and quality check duties. This sub-segment of staff can be the core-trained team for all quality related issues on shift. Regardless of how this is solved and resourced, the ultimate goal should be clear (because you wrote it in your quality manual): Get the quality check done as close to the front line as possible.

As mentioned in Chapter One, as breweries expand there is a history to struggle with "siloing" off quality checks, inflating the quality staff, and not implementing quality at the source. It seems easier because it takes too long to train new staff all the time, or there are too many other new processes being asked of operations. But the silo mentality is a cancer—once it starts, it grows and is unmanageable. Before you know it, you have a quality staff of 25 to keep up with the checks. The policy should be clear to strive to keep the quality staff small, have them focus on establishing standard protocols and assure checks are consistent, and then train the operations. It is a good practice to establish a routine review for what checks are conducted by quality today, and have a plan (maybe a five-year plan) to transition checks. Also, never put a new check in without the long-term thought of who will conduct it, and if it is not at the source where the decision results in process adjustment, when it will be transitioned to quality at the source. Keeping a master list of checks helps manage this process (see Appendix B).

Quality Skills for Brewery Staff
Let's consider the skills a staff member must acquire for a quality-at-the-source program to work well. First is the foundational training. Staff, whether they are brewers, cellar staff, or forklift drivers, should have a quality foundation. Some may be well trained on the entire brewing process, and why process fails, etc. This is not a quality foundation, though. A quality foundation is the knowledge and skills that allow the staff to understand the quality management principles that apply universally to any operation. Foundation skills minimally are:

- How to perform the test and check the data accuracy (e.g., during a pH check, check against a standard first).
- How to calibrate the necessary instruments or alert the quality department if calibration is needed.
- How to maintain the quality check area and instrumentation to working condition.
- Knowledge and understanding of the sanitation process.
- How to look up the specification of the product being checked.
- Basic understanding of statistics on data variation (finding the mean, variance).
- Basic understanding of process variation and control, and reading a control chart (determining what is an acceptable range).
- When and how to react to an out-of-specification condition.
- How to escalate an issue.

A quality subject matter expert or an organization such as the American Society of Quality can help the brewery establish basic foundation training if you don't have it in your brewery.

Maintaining the work area extends into the quality work stations, even if this is in an area on the brewery or cellar floor. This needs to be very clear to the operations managers so they may enforce it. Even in a small brewery, expecting the quality staff to clean up after operations is not an effective use of their time, nor does it provide a clear picture of accountability. Assuming it is clear to the operations team that they are responsible for product quality, then they are also responsible for conducting the checks to the best of their ability, and keeping the check area clean, tidy, and in working order. A way to help oversee this is the use of walk-around audits, or a visual picture at the station to show what "good" looks like.

The Second Line – Middle or Upper Management
One of the best values to build in any size brewery is, "All managers in this brewery are quality managers." If the operation is large enough to have a quality manager on-site, it is widely assumed that they should also manage the personnel conducting every quality check. But this isn't the case, and they generally only manage the lab staff, chemists, and microbiologists. In terms of operations personnel, the quality manager should only be asked to support the operations team for quality check training. The operations managers, in addition to the quality manager, should have a broader skill set in statistics and data interpretation, as it is their duty to summarize how well they are doing in terms of quality process control in their area. In other words, managing personnel that conduct the front-line quality tests, and the results, falls to the brewery operations managers for their respective areas—not solely on the quality manager.

So, what skills should the second line of management have? In addition to the usual brewing, packaging, and shipping process knowledge, they should have a foundation in quality, and they should also have a deeper level of understanding of the maintenance issues that can crop up with the equipment they run. If that is in place, they are ready to take on a middle-level management skill set for quality as well. This is the same list used for anyone in maintenance, or "asset care."

- Foundation skills (previously mentioned)
- Higher level of statistics about process control (SPC charting, identifying trends)
- Principles of problem solving
- People skills (to encourage their staff in results-based decision making)
- Sanitation processes in the entire brewery

Statistics and problem-solving skills are a special mention here. Some personnel are natural problem solvers. They like to get to the root cause of problems and fix the situation. Root cause is the true root of quality issues that, once fixed, removes the risk of repeat failure. This is a good skill to have in a brewery, because processes and equipment break. Unfortunately, sometimes good intent can be more detrimental than helpful, especially if: 1) Root cause was not determined and the "fix" for the problem causes the issue to re-occur or other issues to occur; or 2) The high-priority problems don't get resolved—instead, time and energy are spent on troubleshooting low priority issues, impacting the bottom line.

When problem solving is implemented properly, the high-priority problems are not only well managed, but the brewery should reap the rewards in higher quality products, lower costs, and better efficiency. In addition, the root causes of issues are addressed, saving resources for other issues. For these reasons, natural problem solvers should sharpen their skills and be provided with proper training to learn the problem-solving process. (Problem solving as a strategic component to a quality program will be addressed in more detail in Chapter 5.)

There are general steps to take when tackling a problem. These steps can be used as "stage-gates" in communicating status of solving problems. The steps most commonly referenced are: Define (the problem statement), Measure (select what to measure, by whom, and how), Analyze (analyze the data, and get to the root cause), Improve (choose various ways to improve, and pick the best), and Control (plan to maintain the improvement). This is commonly

Problem-Solving Training

The larger the brewery, the more people you have to communicate with that you are working on a solution to certain problems. The harder the problem is to solve, the longer the steps will take, and the more confusing the communications can become. Therefore, as the brewery sharpens other quality skills, having a practitioner skilled in the art of problem solving can benefit the brewery in simplifying communications and managing expectations of outcomes.

It is important to mention that the skills of the problem-solving expert are very involved. It involves many years of statistical training and practice of complicated problems with lots of data. These subject experts are sometimes qualified with a Six Sigma training level. Not every manager needs a Six Sigma Green Belt (the lowest level of training), and not every brewery needs a Black Belt (the highest level of training) on staff for problem solving, but some sort of quantifiable skill development is recommended.

For this reason, Six Sigma is a great resource to put together a sample training plan. The skills involve not only learning a standard process of Define, Measure, Analyze, Improve, Control (DMAIC), but also a building of manufacturing statistics knowledge (ASQ, 2014). The American Society of Quality conducts certified Six Sigma courses and their website offers excellent background for those considering Six Sigma as a means of bringing process improvement structure to the brewery. More at www.asq.org.

Minitab is an example of a great statistics program to use if problem solving is at a higher level in the brewery. If purchased, Minitab offers online, self-paced statistics training that is excellent (www.minitab.com). (Microsoft Excel has a statistics package, and there are other software programs that run off of Excel data. A statistical package can help explain data via statistics, however, training in running the statistics package is required.)

referred to as DMAIC (da-may-ic). There is a lot written about DMAIC. It is a simple process to follow, and training is available if there is no expert on staff. Even without a formal program in problem solving, though, the management team should know enough about data, statistics, process control, and the problem-solving steps to know when it is a good time to act, when it is not, and what good data-driven decision making looks like.

The Quality Staff
The quality staff usually has a special set of skills that also has to be developed and managed as part of a quality program. The best reference is the American Society of Quality (ASQ) body of knowledge required for *Certification for Quality Technician* and *Quality Manager*. Based in Milwaukee, Wisconsin, ASQ has published several body-of-knowledge documents for quality personnel (technicians and managers). In addition to quality management skills, the brewery quality team must take on a few other skills. Assuming that the foundation of brewery process has been established, and front-line and second-line quality skills are in place, these are additional skills:
- Chemical and microbiological tests for the brewing industry
- Raw material quality control (talk with suppliers, know how their process works), including packaging, cardboard, glass, malt, hops
- Brewing, cellar, and packaging process controls (if automated process controls are in place)
- Quality control statistics
- Calibration and maintenance of test equipment (in-line or offline testing instruments)
- Automation controls and their limitations
- Sanitation processes in the brewery and the chemistry related to the sanitation chemicals

Quality managers must also have the people skills for managing through influence. This is probably one of the most difficult skills to learn. Influence is required because usually the quality manager must not only run a competent staff in the brewery, but also be the person who evolves the brewery's quality over time. They have to choose strategies or methods to work within, and then convince their peers and upper managers or owners of why the strategy would work. It is also highly advisable to begin to speak in business language, not just facts and figures, or Six Sigma statistics, but in dollars and cents. Understanding a balance sheet, how to state the business case, and return on investment goes far in communicating why new initiatives, engineering fixes, or programs should be started.

Assessing Skills and the Organizational Design
Once a year, as planning for the next year's budget is taking hold, it is wise to review the organizational design for the quality team, plug in names and faces, and take a hard look at the skills and skill gaps of the team. Discuss what skills are missing, from the basics to the advanced, and what the path to closing the gaps might look like. If this is done during the budgeting process, the brewery may plan for certain personnel to register for quality management training. If the brewery is small and has one person wearing several hats, including quality management or governance, it is even more important to plan for their skill attainment. That person should be provided the time to build skills in statistics, problem solving, and quality management, and not just focus on staying abreast of brewing science. If separate training is not an option, look to other industries for support. For example, the trainee can mentor with a local food-based quality manager, or join their local American Society for Quality chapter and meet more folks dedicated to the trade in other industries. It is highly encouraged to remove the blinders of being around only brewers when working in quality management. Exploring bigger organizations and how they manage data, problem solving, and other quality related issues is a great way to learn.

CONDUCTING THE RIGHT TESTS AND CONDUCTING THEM WELL

After a plan to assess skills has taken root, the brewing management team can look at the next critical component to a quality program; that is, conducting the right tests, and conducting them well, no matter the size of the facility. The quality tests that a brewery runs are usually summed up into one table as part of the quality manual. This table is called the QA/QC control plan (Appendix B) and these are the measures that ensure the products and the process remains in control. "Metrics" is a broad term that relates to bigger initiatives (such as energy use, cost per barrel, consumer complaints per year, or other), and is not covered in this plan. Usually a brewery can get its control plan started on day one of operation.

A well-trained brewer that understands the basics of beer quality will generally create a control plan that will look very similar to many other breweries, except for the fundamental variation in the individual brewery's process risks. Risks are the threats to product quality that are the result of process variation, raw materials, and controls. Most breweries take on some sort of risks that are additional to the inherent risks in the basic brewing process. For example, a brewery can choose to force carbonate or bottle condition to carbonate their product. These are two very different processes that bring different levels of risk to the ability to control finished product consistency. Breweries that choose bottle conditioning must put in place additional checks and controls that other breweries would not have to do. Bottle conditioning is a choice, and because it increases risks, and requires additional monitoring, it should be inherent in the brand identity. A brand identity may compel a brewery to take on riskier processes, but so will growth. Growing in size and in geographic footprint increases risks and requires a different sampling plan. (See Figure 3.1.)

Differences in product and process complexity, such as yeast varieties in the brewery, expected shelf-life of products, and complexity of raw materials are just a few more examples of why one brewery's quality test plan would differ from another's. Another factor for quality program variation is resources. Some very small breweries simply don't have the personnel to use in a single capacity. Their frequency of sampling and assessing quality may look radically different than a brewery that supports more staff.

Shelf Life and Stability

As part of product specifications and process control, the brewer's QC efforts ultimately shape the shelf life and stability of a product. In a brewery with a long-established program, this is where a lot of time and effort is spent. Quality checks include:
- Reducing oxygen ingress
- Physical stability, including protein or polyphenol hazes and sediment over time
- Haze reduction/control
- Flavor stability, including checking flavor after abuse (e.g., in a hot room)
- Freshness, including consumer feedback

Conducting Risk Assessment

To keep the total tests manageable in terms of workload, a risk assessment should be performed. A risk assessment will incorporate the specific nuances and risks of the brewery's product (ingredients, suppliers, location), processes (special equipment, special processes), and variation (brewery's ability to control the processes). The risk assessment is a formal way to map the brewing process, identify risks, and review the control plan on a routine basis. It isn't necessarily done at the opening of the brewery. Once the brewery has established the basics of governance, has a beginning control plan, and begins to see expansion and new product additions, introducing a standard format to assess risks is advised. (See Figure 3.1.)

Prepare by building the right team and a few people can conduct a risk assessment over several hours. The assessment team should be well versed in the process, the quality plan, and the goals of the brewery's growth. Why conduct a risk assessment? Growth brings new challenges and added risks to product quality. Additionally, changing the product mix with new products, new raw materials, and new suppliers adds complexity. With complexity can come mistakes, missed checks, or, in a worst-case scenario, product pullbacks due to unknown risks that were introduced. To keep the brewery staff sane, a risk assessment brings the actual cost of complexity and requirements for monitoring quality in new products and with growth to the forefront.

Example List of Test Types Conducted in Various Breweries by Size

Brewery Size	Test Type
0–3000 BBL	pH of water, wort, beer, yeast Plato or gravity of wort and finished beer (extract, apparent extract, and specific gravity) Forced wort to test microbiological stability in general Yeast counting Sensory analysis on finished beer (go/no go)
3000–15,000 BBL	As above, plus: Basic microbiology tests using media that is easy to prepare Color Bitterness Titration of cleaners and sanitizer levels Rapid fermentations of wort (yeast fermentable extract) will help determine if the pitch process is the problem, or the mash Alcohol testing Turbidity Microbiological control on standard medias (yeast, water, and beer) Fill levels Simple sensory analysis on raw materials (malt, water, hops)
15,000–150,000 BBL	As above, plus: Free amino nitrogen in wort Basic raw materials test on hops and malt (Hops - α- and β-Acids by spectrophotometry, hop storage index, hop teas, physical analysis, malt physical analysis). These can be provided by the supplier. Simple metals such as calcium, magnesium (in wort, beer or water) Diacetyl (via sensory or other) Dissolved oxygen (in wort, from tanks, in finished beer) Gluten (via rapid tests) Sulfur dioxide Dimethyl sulfide Bottle or can closure analysis Finished packaged good physical analysis Finished packaged good gas analysis (CO_2 and dissolved oxygen) Extensive sensory analysis on finished goods and in-process products, shelf-life testing Calories and nutritional analysis (may be outsourced)
>150,000 BBL	As above, plus: Gas Chromatograph of volatile chemicals (esters, alcohols, diacetyl, VDK) Additional metals such as potassium, iron, zinc, copper (in wort, beer, or water) Possibly iso-alpha-acids in beer and wort by HPLC Possibly yeast differentiation by fluorescence staining, or PCR Possibly ESR testing of raw materials and finished beer

Figure 3.1: This figure is an example of how a brewery can add tests based on the additional risks to product quality that growth will bring. Note that as risks increase the test types become more sophisticated, and the human resources to manage these tests must also become more sophisticated. Some of these tests can be outsourced. Reference the Methods of Analysis (MOAs) published by the American Society of Brewing Chemists (ASBC) for the tests shown in this table.

There are a couple of standard risk assessment tools used to conduct an assessment. The first one to discuss is the Hazard Analysis and Critical Control Points (HACCP). HACCP is essentially a structured risk assessment, with a control plan, that is traditionally scoped for *food safety* only. Food safety risks are those that will impact the public health in some manner if not properly controlled. Risks are categorized as microbiological (pathogens), chemical (such as chemical residues), or physical (such as glass).

THE COMPONENTS OF A QUALITY PROGRAM

Example of New Product Risk Failure

Brewery X decided to introduce a new Belgian-style beer to the market. They brought in new yeast, propagated it separately from their standard yeast, and brought in new raw materials, including rye. The yeast was a slow-fermenting strain and did not fully ferment out all the sugars in the beer. This left a sweet, slightly under-fermented finished product. The brewmaster decided it met the criteria of the brand, though, and the brewery began production. Initial sales were great, however, as the weather warmed the sales slowed. The brewery was sitting on 200 barrels of the finished beer, and moved the beer to a bright beer tank that was tapped only when sales called for it. The beer sat for three weeks, and it was slowly drawn down via packaging. Several weeks after the tank was emptied, the brewery began to receive customer complaints from the field. The beer that came back to the brewery had soured. It turns out that the bright beer tank had a low level *Pediococcus* infection. Because the beer had residual sugar and nitrogen, the microorganisms began to slowly grow in the finished beer, resulting in a highly sour and buttery beer. All the beer had to be pulled back from the market, at a cost of over $150,000. A risk assessment may have picked up on the need for this brewery to fully empty the bright beer tanks, and to alter the mash cycle to allow for complete fermentation of the beer.

This story shows how one new product, incorporated into a standard process that is not ready for the new risks, can result in a costly pullback. New risks become more obvious to recognize with experience. Some to watch for are:

- New yeast and propagation – risk of cross-contamination of current strain
- New raw materials – risk of impacting brewhouse performance, and bringing in new micro-flora
- New products with very different finished beer chemistry – higher risk of infection
- Slow turnover – risk of inefficiency in cellar, and packaging, impacting other product quality

Because HACCP does not assess risks to product quality, it is sometimes hard for those in the brewing industry to comprehend this type of risk assessment and limit the HACCP plan to food safety only. Most quality checks in brewing are quality related, not food-safety related. This is because beer is essentially a relatively low-risk food to the customer. HACCP focuses the scope of the risk assessment to three critical hazards that can harm or injure the final consumer: microbiological, physical, and chemical. There are no microbiological hazards (pathogens) known to grow in beer that can harm or injure the consumer. Therefore, if a brewery has a formal HACCP plan, its focus is on the opportunities in the process for physical and chemical hazards such as broken glass or chemical residues, including mycotoxins from mold. (A cautionary note: Even though pathogens have not shown the ability to grow or stay alive in beer, with the level of innovation today, every brewery should know what raw materials it is bringing in, and if they may contain pathogens. Mycotoxins or beer with fruit can raise these concerns. If that is the case, ensure those materials are being thermally treated or otherwise contained.)

Some breweries have used the HACCP philosophy and structure to create a risk assessment for product quality as well as product safety. This is a great tool for using risk assessment and can be extended into product quality. However, keep the actual HACCP plan separate for audit purposes. HACCP as a tool is not too hard to grasp, and most can conduct this type of assessment of their process with little or no training, just a bit of reading and thought. The HACCP process, in its entirety, takes the results of the risk assessment and structures a food safety control plan to control the risks (what test, by whom, how frequently, and what the corrective action is). Examples of HACCP risk assessment can be found in Appendix C.

> **Hazardous Critical Control Points**
>
> HACCP is technically a food safety program. Pillsbury introduced the program per the request of NASA during the 1960s. NASA wanted a rock solid risk review and control program to prevent any food borne illness in space. HACCP is not required for breweries operating in the United States as of the date of this publication, however, a risk-based assessment for food safety is recommended as part of the FDA's Food Safety Modernization Act of 2011. For more information on HACCP in breweries, the Master Brewers Association or Beer Canada's websites are extremely helpful.

Another risk-assessment tool that can be used to structure your thoughts is a Failure Modes and Effect Analysis, or FMEA. An FMEA is a traditional risk assessment that establishes a quantitative score for each risk to product quality in your process, in order to prioritize areas for improvement. An example of an FMEA-type template is located in Appendix D to help get you started.

Conducting a Risk Assessment with FMEA
1. Write down all your products and what they contain (raw materials).
2. Map your process to reflect the production of all products.
    a. Don't get too detailed, as there is no need.
    b. Keep the steps of the process at a higher level, as if you are explaining the process to someone who doesn't know how to brew. Keep it very simple.
    c. Write a brief explanation of the process.
3. Identify how the process would FAIL quality—this is the "failure mode."
    a. EXAMPLE: Mashing in failure modes
        i. Too hot/cold temperature mash water
        ii. Paddle speed too fast
        iii. Bridging of grain
4. Identify process inputs and output controls that are monitored at each step to prevent the failure mode.
    a. EXAMPLE: Mashing in
        i. The process inputs are mash water temperatures, grain/grist flow, paddle speed, etc.
        ii. The process outputs are a visual on grist consistency and the test of starch conversion with iodine.
    b. Detail if there are additional process controls for different products.
        i. From our example, if rye is used in a product and it can bridge on graining-in, note that as an additional risk. The control may be visual monitoring.
5. After inputs and output controls are identified, determine on a scale of 1-10 (with 1 low and 10 high):
    a. How risky to product quality is it if the process fails? (severity)
    b. How likely is it that the process will fail? (occurrence/likelihood)
    c. How good are the quality checks at detecting and controlling a process failure? (detection)
6. Multiply the severity, occurrence, and detection numbers together and you have a risk priority number (RPN).
    a. The RPN, if really high, should be addressed.
    b. Detection is usually the critical factor to fix. This helps inform those responsible for quality governance what to address in the quality system.

Using this approach captures *what* each check is and *why* it is being done. The maintenance team will find value in understanding the importance of the different mechanical systems, and many quality failures in the brewery can be attributed to mechanical failures of some sort. New staff will also benefit by using the analysis as a training document for reviewing the process. It can be used in the same vein as HACCP, which focuses on chemical, physical, or microbiological hazards that are critical to customer health. FMEA assessment can easily be conducted for every new process coming online, therefore adding to the quality system as required. It does take time to conduct a full risk assessment to this detail level in a brewery. It also requires alignment with the brewery teams to conduct the assessment, and to agree on the risk levels. A living document like this must be maintained.

If FMEA- or HACCP-style risk assessment is too much structure and formality for the brewery, the minimum you should do is map the process, and write down the controls you have (inputs/outputs). Summarize these controls in a control plan as noted in Appendix B. Additionally, Appendix E shows an HACCP process map that could be modified to show quality control points if necessary. The upside to the process map approach is that it is quick and easy and it captures the quality system in a clear and visual manner.

The downside is that it misses addressing failure risk. Assuming a risk review was conducted, and the correct monitoring program or control plan has been summarized in some standard format as illustrated in the appendixes, the next key objective is to determine the limits to each check.

Setting Specifications

Once the brewery is satisfied that it has selected the right tests from its risk assessment, it must go about the duty of setting specifications, or limits, for each check. "Specifications" are the set quantifiable limits on the process or product parameters based on customer requirements. "Control limits" are statistically set, hopefully under the specifications, and should require reaction because a process or product is trending to exceed a specification. Breweries tend to be "metric dense," or heavy in data and data acquisition. We measure a lot in breweries. The term "spec" is also thrown around loosely, sometimes meaning a customer requirement, sometimes meaning a control limit for the process. Therefore, specification setting can become complex because, effectively, we have to set specifications *and* process control limits in tandem. It isn't recommended to make this process any more complex, however, setting specifications *and* the control limits requires careful thought and a standard approach.

The approach recommended here is fundamental to setting specifications and is easy to implement no matter the size of the brewery. It is based on simple rules.

Some rules for setting specifications:
1. Set specification limits using customer (end user) requirements, and use standard control charting to set control limits.
2. Prioritize your risks and your key control parameters.
3. Know exactly what to do if a specification or control limit is breached. Have a plan for both that is not over reactive or costly.

If you already have specifications and control limits in place, check them against the criteria here to be sure they are not overly complex. Overly complex specifications can cause delays in product releases due to confusion or needed clarifications. Cumbersome red tape, communications, and documentation cause delays as well. Using process control measures that are too strict can also lead to loss of product quality and loss of operational efficiency if the specs merit further checks or require lots of blending.

The first step in setting a specification is the use of customer-based requirements to set specifications, and process control to set control limits. Beer and breweries are famous for getting into a pickle with specifications that are too tight, and too much reaction to control limits. Bitterness values are a great example. If the customer cannot determine the difference between a beer's bitterness unit (BU) level 30, when it is actually 36, then don't set the SPECIFICATION any lower. Hold the specification where the customer would NOTICE a difference. Of course, you may have control limits that are tighter than the specification and also cause a reaction. Figure 3.2 has an example of control limits and specification chart. Control limits are defined statistically (we will talk though control limits in a little bit), but instead of control limits, an arbitrary "tolerance" or "reaction limit" may be assigned to a process. Why the difference? Remember: A specification breach needs to be considered as extremely serious and all hands need to be on deck to contain it. A control limit breach means the brewery needs to react, but not necessarily to the same pains of breaching a specification. If the control limit happens to be close to the specification, the brewery has an issue of process control. They will be reacting

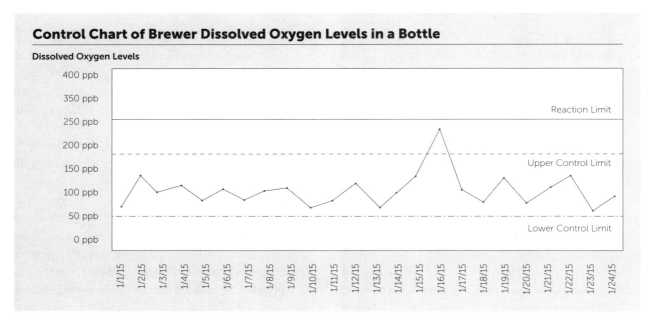

Figure 3.2: The illustration shows a control chart—a statistical tool to graphically represent the quality data. Like goal posts in football, you may want to react when the kicker is always bending the kick to the right and getting close to missing the goal. But if your goal posts were too tight, he wouldn't be scoring at all. Note, the rejection limit should not be lower than the UCL.

to data all the time, because the process doesn't have the capability to stay within specification. Sometimes, though, the brewery has exceptional control. In this case, a reaction or tolerance limit can be set instead of managing data with a control chart. Or, it is possible the brewery has no data yet on a process. Therefore, a tolerance or reaction limit may be arbitrarily set within reason. (See Figure 3.2.)

To illustrate the difference in reaction, let's assume the desired bitterness level in a specific beer is 30 BUs. Our specification is 24–36 BUs, though, because we know the customer will actually perceive a difference at this level. We may have control limits at 26–34 BUs that were set arbitrarily or by a statistic (discussed below). We react to both if the control limits or specification are breached, but the reaction to each is very different. If a test exceeds a specification, then we may hold the batch and maybe blend it with another. If a test exceeds a control limit, we may instead put a watch on that hop type, re-test the hop alpha acids for the lot, or review the data of the batch to see what may have gone wrong. Too many control limits being breached for one test or beer type means there is another level of reaction needed. There may be troubleshooting of the process, adjusting the recipe if root cause requires it, or checking the test itself. The point is, do not hold, taste, or blend unless you exceed a customer-derived specification. This is costly and may actually cause more complaints if not managed well. Excessive reaction to control limits should not be the norm.

On the other hand, no reaction to control limits is equally as harmful. If a specification is going to be breached because of a trending data point, you need to react before you start destroying beer. Ignoring trends, especially for critical variables that impact customer interest (color, flavor, bitterness, foam, or alcohol), will equate to specifications eventually being breached, and customers being lost. Just because you don't have full understanding of the trend does not mean you shouldn't react. For example, an intended sour flavor that may be a result of a microbiological variable that slowly changes or creeps to an intolerable level isn't acceptable. Customers have plenty of other options to try, and less tolerance than ever to a scope creep in a beer. Instead, have a plan to absolutely hold firm on releasing anything that exceeds specifications, even in flavor, and always react to trends before it costs you a customer. Even if you can't determine the root cause, the brewer's best tool is the ability to blend, if set up for it, and continue to create the flavors they want. Use it if trending is causing a beer to no longer be what the customer expects.

Lastly, trending data is helpful to other teams in the brewery. Don't limit it to just a few sets of eyes. Monitoring seam data in cans or dissolved oxygen in final beer are good examples of how control limits are a great tool for the maintenance team and other supporting teams to react to. Can-seaming trends may indicate higher wear on the tooling. If a part is failing before it should, this data helps in discussions with the tooling manufacturer and this may even equate to a recouped savings for the brewery if the tooling is replaced with a discount or refund. (For more on control charting see the references section at the end of the book.)

Setting Limits

There are a couple ways to set limits for control. As mentioned earlier, sometimes control limits are set arbitrarily and are called reaction or tolerance limits. As long as the limits are within the specification, this can be a sound strategy. Reasons for using this strategy may include keeping different beers from being too close together in alcohol or color, or when a new process is just getting started and is in the phase of pre-control. However, if a limit is in place without understanding of *process capability*, it is likely that overreaction will mean wasted work and reduced product quality. It is important to understand this as a risk in setting limits in an arbitrary fashion. Instead, it is highly recommended to set limits using a more science-based approach.

Process capability is a statistic that explains how much process variation one can expect. Every process in manufacturing has a natural amount of variation in its ability to perform its function. Process capability is a measurement of how much the process naturally varies due to natural causes (or *natural cause variation*). Without knowing or understanding the process capability to control the desired range, you run the risk of making adjustments that overcompensate and actually cause *more variation* in the parameter you're trying to control. Deming was first to illustrate this in a famous funnel and bull's eye experiment. This is basic statistical premise, but it can easily be forgotten (Castillo, 2002, 29).

For example, a wort chiller's function is to chill the wort to a certain temperature. Every knockout should hit the exact same temperature if process is in perfect control. Because the desired temperature range on the output of a wort cooler is critical for yeast to ferment at the proper pace, it seems wise to put strict specification limits (upper or lower) on the temperature in which the wort is allowed to cool. However, if the natural variation in wort flow, water flow, glycol, or the temperature of the receiving fermentation tanks causes the knockout temperature to vary more than the limits set, you run the risk of overcompensation, and likely higher variation of the end-wort temperature, causing a cascade of inaccurate flavor variances downstream. There are other parameters that cause the wort cooler to not function optimally, but it may be obvious. These are called "assignable causes." An assignable cause in the wort cooler could be an ice dam that caused excessively hot wort to be transferred to the fermentation tanks. The root cause to assignable cause variation can usually be more easily determined and prevented from occurring again. Higher forms of study such as Six Sigma utilize statistical tools and problem-solving techniques to reduce unassignable cause variation reduction. As a brewery understands its process capability, its limit setting becomes much more robust and helpful in containing real quality control issues.

Variation Reduction

For those who have had some Six Sigma training, you likely know and are familiar with utilizing process capability to set control limits. However, this may be foreign to those who have not. Six Sigma training allows the process owners to dive deep into natural cause variation and determine what is controllable and can therefore help reduce variation even more. The glycol or cold water flow consistency in the wort chiller example may be something the team could work on to reduce hot spots during cooling, and therefore help continually maintain a controlled process (with specifications always hitting the mark). Reducing variation is a cost-saving measure, because eventually the brewer will not be reacting to control limit issues as illustrated in Figure 3.3.

Variation Reduction

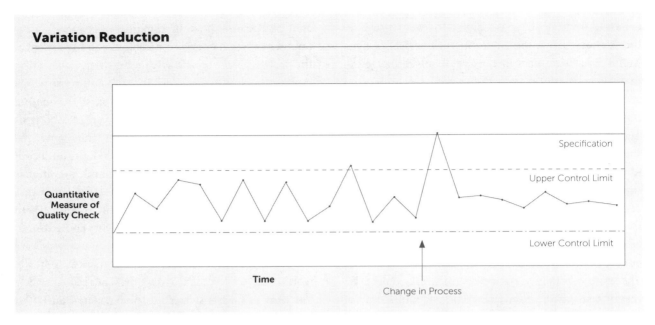

Figure 3.3: Control Chart with variation reduction

So, how are control limits set? If you have a very manual process that does not rely on automatic process controllers, you will likely have to take a few extra steps to make sure you understand the natural process limits and do not over adjust. A quality professional would call this the pre-control phase. This is the one time in which arbitrary or best guess control limits are warranted, as mentioned before.

1. Start with the desired finished quality parameters you want to monitor that are indirectly impacted by the process. In the case of wort cooling these are the fermentation results of alcohol, flavor, and fermentation rate.
2. Determine what the ideal targets of the process controls would be. For wort cooling, we are directly controlling temperature as an output. Ideal control may be +/- 2°F.
3. Lastly, look at the process control inputs and determine whether you can monitor these. For wort cooling, the rate of flow of glycol or cooling water, for example.

If you don't have access to input data, that is all right, however, you may be limited in controlling variation until you can measure the inputs adequately. In this case, we will focus on the output temperature of the wort. We would decide what wort temperature would be considered *significantly out of control* and require correction (such as too hot or too cold), and what temperatures would simply require extra monitoring if you reach them. Then we would allow the process to run in its natural state (don't correct unless it is critical) for 20–30 knockouts. Set a limit on how many batches you will allow that would interfere for critical out-of-control conditions (3–4) during this test phase. Once you have collected the data you can plot it in a control chart—assuming no major quality corrections were needed, the quality indicators (flavor, alcohol, rate of fermentation) stayed within specification, and no major assignable causes of variation impacted the test. You can run a standard Excel or statistical package to calculate the actual control limits. (For control limit calculations, see Swanson, 1995).

This initial set of data represents the natural process variation for this process. This variation can be calculated into a statistic, the sigma level, for the process capability, or *Cpk*. This number is based on the specification targets you set. So, if your process is not capable of meeting the specification due to natural cause variation, you will know it. The lower the sigma level, the worse your process is to meet specification. (The process improvement program, Six Sigma, is actually named after this statistic.) If a process is running at a Six Sigma level, it is highly capable of running within the process target or specification.

For some, calculating a statistic isn't too important. If that is the case, using standard control limits can determine the natural variation of whether a process is too high by visually reviewing the charts. If you want to reduce variation, and don't have more advanced training, use some caution. Remember: If you overcorrect or overadjust a process that is running within its natural capability, you run the high likelihood that you will cause more variation and not less. Use logic and reason to troubleshoot in these cases. Challenge only one input parameter at a time, measure the results, and make a data-driven decision if the change you inputted really helped control natural variation or lower your average. Take the time to learn the natural process variation before attempting to improve variation or lower the average.

Lastly, we must address measurement control in the context of process controls. If you are taking an offline measurement, such as pH or gravity, you cannot set any reaction limit *tighter* than your measurement precision. For example, if you are utilizing hydrometers on the brew deck to control gravity measurements, recognize that a natural measurement variation of +/- 0.2°P is fully expected (check your hydrometer manufacturer's precision scale). Therefore, setting limits at anything less than +/- 0.3°F is asking for more problems. If your target was 10.2°P, and you hit 10.0°P and adjusted the next batch, assume you caused more variation in your process by adjusting the next batch than would be present if you hadn't. In other words, hands off the controls if the process shows variation within the measurement instrument's precision range. Know your instrument's precision for offline measures before setting a specification or reacting to process control limits.

Setting the Frequency of the Check

After process limits are set, one must determine the frequency needed for the quality control check. The frequency of the check needs to be set so the staff can see the peaks and valleys that are expected in a process. This is to ensure variation isn't getting out of control, as well as ensure no process or product is out of specification. The frequency of a check depends on many variables:

1. The initial risk assessment (likelihood the process will fail).
2. The natural process variation. If a process is in control it shouldn't fail too frequently, and therefore any variation of the output should be fairly well maintained.
3. If process input variables are being monitored and controlled.
4. Ease of access to the measurement.
5. Cost and difficulty of the check.

As we have already addressed numbers 1 and 2 on the list above, let's focus on how measuring inputs, ease of access, and the difficulty of the test also establish the frequency of a check.

Quality control evolution in any brewery is to push the brewery from monitoring, measuring, and controlling the outputs of a process to monitoring, measuring, and controlling the inputs. If a brewery has a rigorous process input check on a computer controlled system, the likelihood that an output will fail is slim. This is a fundamental principle to understand. Control a process's input, and the output and subsequent beer quality will most likely not be compromised. For example, controlling the temperature of mash water helps ensure the conversion of starch to sugar. Some breweries have enough control over process inputs that they eliminate or reduce the checks of the outputs, thus saving the brewery time and money and impacting frequency of checks. Back to the mash example, some breweries don't bother to check iodine conversion at the end of every mash cycle. Their process inputs are so well monitored and controlled by a computer that any trending or variation results in adjustments that fix the process before it impacts the result. They monitor the output of iodine conversion only to validate this control. This is why breweries that spend extra money on computerized process control can improve beer quality and reduce time to detect and correct an issue, cutting costs for quality monitoring with people.

Smaller breweries may not have the luxury to put in a highly computerized control system. They may have more manual valves, temperature controls, etc., and only be able to monitor the process outputs. For the mash process example, the iodine conversion is the output to test. It is likely the brewery completes the starch conversion and is able to move the batch to the lauter tun. However, if the iodine test fails, this becomes a "no go" check. The brewery now must document the failure, and then adjust the mash before heading to the lauter tun. This is still a valid check, and one not to drop. In the absence of good input control, keep the output controls intact.

Lastly, the frequency of checking inputs, assuming they are computer monitored, is typically data limiting. If a computer can measure a temperature every ¼ second, and record it, then why not allow that level of control to occur? However, usually a brewery has only so much space to save data. For that reason, decisions have to be made as to what level of control is needed for each process, and what is critical to collect and save versus control and dump the data.

Control a process's input, and the output and subsequent beer quality will most likely not be compromised.

The frequency of checking outputs, in the absence of any input control, should be at the end of every batch process. But here's where it gets tricky. Can the brewery manager assume the one small sample of mash, randomly pulled from the mash tun, reflects the

Sampling for Microbiology Checks

Microbiology checks are almost always conducted offline and in the lab. The results take days to get back, and by then the beer has usually moved somewhere else. These are poor examples of quality at the source, but they have to be done. In brewing, a grab sample is typically taken from a sampling valve as in Figure 3.4. This one point is small, fixed on a tank, and can seem like sampling only from it doesn't fully represent the entire tank. This is because microorganisms may "hang-up" in a tank, in shadows, or under a biofilm. But a liquid, such as beer, has an advantage in being a liquid and therefore homogenization is assumed (though not always the case), especially for the standard quality check.

The question of how much to plate depends on how much risk you are willing to live with. If one organism in 100 mL is too much risk (this is a low level of contamination), then plate 100 mL of liquid. If one microorganism in 1 mL is what you want to react to (a high level of contamination), then plate 1 mL only.

For more detailed information on microbiological criteria and sampling plans, see the reference for the International Commission for Microbiological Specifications in Food (ICMSF).

One last comment: The reporting of microbiological results is in colony forming units (CFU) per level of volume (CFUs/mL, for example). Colony forming units are used because one organism may not have caused the colony to grow on the plate, and instead many organisms may have caused the growth. Additionally, the report *should not* state zero CFUs/mL if nothing grew on the plate. This is a common error, and one that seems to be frequently misunderstood. The report *should* state <1 CFU/mL. Because you are talking microorganisms, there could feasibly be less than 1 in a mL. If you plated a mL, though, this is the most accurate way to report.

Figure 3.4: Sampling Valve. Image courtesy of Perlick Corporation.

entire batch conversion? The standard rule of thumb is that the more frequent the measure the better the data. However, go/no-go checks become a bit difficult to manage. Take five samples, measure five times, and if one is a no go, then what? This is when using one's best judgment comes into play. Back to our example, assuming the likelihood that there is a cold spot in the mash tun, or poor circulation in an area, then failing the check is unlikely. Being overzealous in sample taking is a waste of time and labor. However, if the process has failed in the past, or shows high variation, then it may be worth taking a few extra samples. Generally speaking, though, because beer is a liquid, only one sample is needed to represent the entire batch for go/no-go sensory or analytical chemistry tests. When in doubt, check the process in the worst-case position. For example, a mash tun's worst-case position is where the mash does not get good circulation and may result in "cold spots."

Ease of Access to Data
Breweries have many ways to get information from a mostly liquid process: offline sampling, and taking the sample to a lab or bench; in-line testing using a roving instrument like for CO_2 or dissolved oxygen; or static in-line quality checks with instruments in the pipe.

Static in-line measurements or inputs give the brewer the best tool to see the process unfolding, without waste of products or human intervention. These tests can be monitored the most frequently without an additional cost in human resources; however, in-line monitoring is expensive to install and sometimes to maintain. Taking samples from a tank or pipe and bringing the sample to a station to measure it also costs money in personnel and time. There is the additional slight loss of product in this case. All of the costs for offline testing should be considered and compared to routine in-line testing. If in-line tests provide better data more frequently, and can be cost-justified, it is preferable to use them. Overly frequent offline checks may be limited simply because of time and money.

When it comes to sampling frequency in packaging, you are no longer working with a batch system, and in-line testing is not an option. Instead, every bottle and can represents an opportunity for a quality failure. If you are running a controlled process, testing every can or bottle is not necessary. How does one determine the frequency of packaging checks? This is where a good control charting process is helpful. Let's assume a valve on the packaging line is failing, and occasionally results in high dissolved oxygen in a can or bottle. The more beer you run, the more failures created, but it may still be difficult to find. Is it worth pulling more samples off the line to test for something that may happen infrequently? Likely not. Sampling at a high rate when failure of a valve is infrequent is a waste of money and time. So, the simple rule is: Take the risk you are willing to take and balance risk with reward. Many breweries sample at the beginning and end of a run to check dissolved oxygen and CO_2. Chart these results to keep track of and react to trends. Using this method, you will find the catastrophic failures, and control the smaller failures as they begin to impact a larger amount of product. Trending data on a control chart is exceptionally helpful to find a slowly decaying part. If the maintenance program is robust, the frequency of checks can be reduced. If you have a risk of failure on any one valve, sample them all every shift. If you don't, then randomly sample a couple of cans or bottles. The only exception is can seaming. Sample every seam head at least once each shift or packaging run to ensure there are no issues with the seam. Plotting this data and reacting to trends is also required in seam checks.

Lastly, an "acceptance sampling plan" can be used when a large lot is on hold because you detected a large failure (high dissolved oxygen, for example). In this case, the lot is split up into smaller lots and a sampling plan is constructed. The user of a sampling plan understands there is a level of confidence they have to assume. As only a small proportion of the lot is sampled, there is still a chance that the sampling plan will not catch the defect. This "acceptance quality level" is reported as a percentage, and it is how many non-conforming units the brewery is willing to send out (usually 1% is selected). The web has many free, simple calculators that allow a user to punch in variables and allow the calculator to determine how many samples to take per lot.[1] Acceptance sampling plans for regular testing are

---
[1] http://www.sqconline.com/squeglia-zero-based-acceptance-sampling-plan-calculator

not recommended for a brewery that is running properly and has well-maintained equipment for standard quality control checks. Use acceptance sampling plans only when a defect has been detected and the lot has to be assessed, go or no-go.

The frequency of checks may come down to simple math and cost control. Some checks are important but are very time consuming or difficult to conduct. If a check requires human interface to take a sample and bring it to the lab (as is the case for microbiology checks), there will be a natural limit to this simply because the brewery cannot afford to hire a large sampling staff. If a check is important, but very difficult to conduct, consider the risk assessment, the natural process variation, and whether inputs are controlled already, and then determine if the check is necessary. The cost benefit of difficult or expensive tests may impact the frequency and the variety of strategies used.

ASSURING THE TEST RESULTS

Conducting a test well means having a quality assurance program that assures the test result is reliable. How is this done and by whom? Quality assurance tasks include anything from calibration, checking a test against a standard, and testing the repeatability or reproducibility of a test result. Quality assurance procedures also entail maintenance of the test equipment.

A quality control plan without a quality assurance plan will not be successful, because if QC test results are constantly questioned as not "good" results, they are not acted upon and thus quality of the product suffers. There is an issue with QA when a brewery staff doesn't react to a quality control measure because they "don't believe that number." A quality assurance plan ensures there is no debate on the measurement results. The quality control plan in Appendix B has a corresponding quality assurance plan. Note that the plan will differ in every brewery because the complexity or simplicity of the tools used to measure processes will differ.

As we have mentioned before, the people who do the job should be considered first. You will need to decide who will conduct the quality assurance tests, but also who should oversee the program. In the best breweries, there is one person or a single team who is assigned to look at the entire quality control plan and determine what the quality assurance tasks are for each test. This can be a management or governance role. Having a single overseer of the plan helps ensure some consistency and also reduces debate. It is easy to cut corners on quality in general, but quality assurance is one of the easiest areas to ignore unless you have assigned the duty of overseeing and analyzing the QA plan.

Once someone has reviewed all QC checks and the corresponding QA test plan, within the QA plan there will be both difficult and easy tasks to perform. Easy, basic tasks can be delegated to the operations team. For example, the brewery staff conducting the quality check can perform a calibration or check a test against a standard. A pH calibration check would be dependent, of course, on the equipment (if it is in-line or offline, if the probe requires a lot of maintenance, etc.). The pH of the wort or finished beer is a very important test, but if the pH probe is dirty or the equipment isn't calibrated, an inaccurate result may end up causing more issues downstream, not to mention costly decisions. If it is an offline probe with standard care and maintenance requirements, it is recommended to make calibration part of the standard duties for the staff conducting the QC check. Asking a quality technician or laboratory technician to calibrate a pH probe daily, instead of the staff using the pH probe, increases the chance that the staff won't "believe the data" and subsequently act on the QC data. For consistency in quality assurance with any QC check, proper training is necessary on how the test works, how to check the test against a standard, and when to calibrate the equipment. As part of standard procedure, if a result does exceed a quality specification, the staff member or laboratory technician should be able to explain what was done to re-check to ensure the test is giving an accurate result. For the tests that have critical importance on go/no-go for a tank release, such as diacetyl or alcohol content, debating the test results should never happen. Proficiency testing—that is, testing the lab's results against another lab (via ASBC or other means)—is a good way to check the lab's accuracy. Put quality assurance as close to the staff as possible, and then focus the energy of the brewery on rectifying the tank or beer at risk, not debating the test results.

The more difficult quality assurance job duties involve some instrumentation expertise. Complicated laboratory equipment, such as gas chromatographs or any in-line test equipment for gravity, pH, oxygen, and offline test equipment for gas testing or gravity checks, require annual maintenance and sometimes more difficult calibrations. These should be conducted by someone trained in the instrumentation. If the check is an in-line instrument, knowing how the instrument corresponds and relates to any controls is extremely important prior to doing any calibrations. If you have to bring in an outside contractor to assist, be sure that skilled and knowledgeable operations staff are included in the work and are part of the communications. Routinely, outside expertise should be brought in to calibrate scales and pipettes, and to check other instrumentation, such as temperature controls, on a scheduled maintenance routine.

The last component of a solid quality control program is having a standard repeatable way to correct any out-of-specification conditions in finished beer or any process that is out of control. This is called "corrective action." Corrective actions should be:

1. Documented
2. Monitored for trends
3. Improved upon

If a corrective action is needed every time a batch is run, then clearly something must be done to review what is happening or determine if control limits need to be reconsidered. Assuming this is not the case (and it shouldn't be), documenting what happened, and what was done, should be a fairly simple process. Standardizing not only how the action is documented but also what should be done is highly recommended. Every out-of-spec condition cannot be predicted, so assume you can capture 50–75% of the corrective actions on your control plan. Other conditions you can't predict should have enough flexibility that the staff can still *contain* an affected batch or lot for further quality evaluation. Using a template to document the action is recommended. It should capture what the out-of-spec condition was, what was done and who did it, and the result. This document becomes part of the quality program and is reviewed to determine areas for improvement. If one process or one style of beer requires too many corrections, then the operations and quality teams can look at product design, process controls, or even measurement control as areas in need of improvement. Have a routine review of corrective actions with upper management as part of a standing quality review. (Many examples of corrective action reports can be found on the internet.)

In summary, the heart of a quality program requires some extensive thought, diligence on setting limits, and determining the actions to take if the limits are exceeded. Quality programs that are constructed in this manner have a better shot at the discipline it takes to manage a brewery's quality output. Invest the time and reap the rewards.

KEY TAKEAWAYS

1000–15,000 BBL Brewery	15,000–150,000 BBL Brewery	150,000+ BBL Brewery
• Set rejection limits and begin monitoring the process • A very basic QA plan should be in place • Document a high-level risk assessment of food safety risks and how they are controlled	• Understand the control limits of critical processes, corrective actions become more programmed • QA becomes more formal and in-depth • Risk assessment becomes more deeply entrenched in the regular review of the quality system and food safety systems • Risk assessment added to new products	• Control limits are actively being worked to reduce variation • Corrective actions begin to feed into a formal problem-solving program • QA duties begin to be transitioned to operations • Risk assessment is part of standard operations review • Formal risk assessment audit may take place

FOUR

SUPPORTING FUNCTIONS TO THE QUALITY PROGRAM

So far, we have focused on quality management and operations teams, and how they support one another in governing and managing product quality. Quality is the job of every employee in the brewery and flows top down from the example set by management. From the warehouse employees and forklift drivers to the "feet on the street" sales team that should be monitoring quality on the distributor, retailer, and consumer end. It truly takes every single employee in every single position to ensure that a quality product is produced and distributed. In this chapter we will turn our attention to other areas of an organization that support the overall quality program and the product's quality. These areas include human resources, asset care of equipment (also known as the maintenance department), sanitation, and record keeping (including complaint records). These departments help brewing, packaging, and shipping operations function more smoothly in many ways. From hiring and firing of personnel to managing a maintenance plan, these functions can be critical to the operations teams. They can also be critical to the brewery's ability to produce a quality product. The processes and policies in the supporting functions bring another layer of proactive support to the quality system. While the quality control plan focuses on catching quality issues, the prevention of quality issues happens via well-orchestrated policies and procedures across many departments and teams. Let's explore why.

HUMAN RESOURCES

Everything that is accomplished in the name of quality happens through good management of human resources. Mentioned previously, skills management is a foundation of building the quality at the source program, but it is also critical for any process, whether it is learning to properly clean and sanitize a tank, to change out a pump seal, or to safely drive a forklift. Managing the skill development (or learning development) process for the entire brewery can be a large task depending on brewery size. Even in smaller breweries, bringing attention to how people learn can be difficult in the wake of production demands. The objective in skills management or learning and development (L&D) is to develop training plans, standardize the look and delivery of training, and organize training so the brewery operations managers are well supported. Many times this function is delegated to the operations staff as a sub-part of a production manager's job, or a separate person may act in this capacity in larger facilities (the learning and development manager). Whoever has the responsibility, L&D managers have to consider what skills are needed for every position and how they will deliver the necessary training. For this, they need the input of many other managers to help them put a plan in place.

As in setting up a quality program, it is good to first lay out who is responsible for the oversight of the training program. Many smaller breweries don't have a training manager and simply get through the week, training operator to operator. Training falls solely on the operations manager to oversee.

In mid-size breweries, human resources may oversee some training, planning, and standardization, and operations managers or the quality department oversee the rest. Larger breweries delegate the actual training, planning, and standardization to a single person or team. They plan everything from safety to quality, and even provide culture and leadership training for every position.

How to know when to add a focused training manager is dependent on the brewery and the complexity of tasks. A good rule of thumb is that once a brewery is large enough to support a brewery staff of more than 10 people, and has at least one layer of management and teams, it should be evaluated if the management of learning and development can be delegated to one person. Whoever manages learning and development needs to know the entirety of the business, including the complexity of the job functions. Specifically, the specialized skills needed in quality, safety, and asset care or maintenance must be well understood. Ideally, those who oversee the training program have the subject matter expertise in some area of the brewery and know the inner workings of the brewery. Because they won't know every job and skill, they should be empowered to pull in the appropriate expertise either within the brewery or outside of it to create a complete training program. A complete program includes basic orientation, all aspects of quality foundations and management, safety policies and procedures, and process skills for every position.

Once you determine who oversees the L&D management, you'll need to tackle exactly how you will deliver the training. Most breweries rely on operator-to-operator training for their employees. The team element of peer training can be the most efficient and rewarding for both trainee and trainer. Other training might be done in a classroom, or in a pilot brewery for more hands-on learning. A mix of reading, listening, and hands-on training is best for most adults in a manufacturing environment. When thinking through the *how* to train, be creative. The most impactful training is fun and engaging, broadens how the learner thinks, and allows the learner to ask questions and take risks. What does this have to do with product quality? Everything, because the quality of the product leaving the brewery is reliant on operators being well trained in their duties. But they also should not feel pressured, overwhelmed, or afraid to ask questions or raise concerns if quality is questionable. Good training will build the culture of speaking up when needed.

Another resource for training can be the brewery's vendors. Your suppliers of chemicals, malt, and hops may have a specific training session they can invite brewery staff to, and may even bring it directly to the brewery. They are a great resource.

Finally, you need to consider what training each employee needs. The list is long and varied in a brewery. Brewery staff need skills in not only the brewing process (process training), but also in the supporting jobs, including sanitation, quality checks, safety, and how to find the rest rooms. This results in an immense list, depending on the job. Add in the fact that brewing process training should cover not only what the process is and how it is done, but also why it is done; one can see it takes years to really learn all aspects of a job in a brewery. The best breweries will have a basic, an intermediate, and an advanced set of process knowledge or skills identified for every job duty. If starting with a blank piece of paper, it will require those responsible for operations to list every process or procedure their team members perform. After creating the list, categorize the process as having basic, intermediate, or advanced levels. Lastly, decide how to sequence the processes in order to learn the skill most efficiently. Bundling many processes together as basic, intermediate, or advanced can be a convenient way to determine if an operator is ready for the next set of learning objectives in another area of the brewery. It can also be the basis for salary increases and job title progression. See Table 4.1 for a sample list of skills for a Laboratory Technician.

Quality Services – Brewery Laboratory Technician

	Basic Skills	**Intermediate Skills**	**Advanced Skills**
Math	Dilutions, blending formulas, ratio math	Mixing chemicals, Molar/Normal conversion, basic statistics	Advanced statistics
Measurement	Pipetting, all wet chemistry measurement tools	Quantitative analysis	Non-linear calibration
Microbiology	Plating (pour, spread, membrane filter), media preparation	Yeast assays, anaerobic chamber tests	Advanced assays such as Polymerase Chain Reaction (PCR), a test to identify the microorganisms via genetic markers
Chemistry	pH meters, spectrophotometer	Gas chromatograph	Advanced instrumentation
Sensory	Basic – Go/no-go		Master sensory – descriptive profiles
Instrumentation	Calibration of instruments online, offline	Calibration and adjustment of all measurement equipment	Maintenance of small pumps, valves, and instruments

Figure 4.1: Learning matrix – This figure shows a simple learning matrix for a laboratory technician. They may need to prove their skills before moving on to the next level of skill acquisition. Sharing a basic matrix such as this allows personal responsibility for their own learning and development, and lets them see the path to advancement.

Once the skill sets for different areas of the brewery are catalogued, someone can be assigned to create the learning aids that help with the training. These learning aids may be a one-page reference sheet that explains the what, how, and why of the process, or detail step-by-step procedures. These extra training aids support the employee through their training. When developing learning aids, keep in mind that employees must learn what processes they will perform, how to perform those processes, and why they will be performing them. Leave any of those questions unanswered and the results are typically mistakes, or worse, quality holds. Answering what processes a person must know, how the processes are performed, and why the processes are performed for every job function reduces the unanswered questions.

This is a good time to mention standard operations procedures (SOPs). Usually operator-to-operator training will exhibit the standard way of conducting a process. For example, "Hook up this hose to this valve, turn the valve open here, then purge the liquid like this." These steps may or may not be written down, but they are considered SOPs. If the procedure is not written step by step, there is a chance the new trainee will need more time and practice with another peer worker. There is also a chance that bad habits and non-standard procedures will be passed down from one operator to another, possibly impacting quality of the product. A good example of this would be a maintenance duty in which the wrong grease or lube is used. If a food-grade lube must be used in certain areas of the brewery, then make sure the training stipulates this requirement and document it accordingly. Having a set SOP for every process, preferably written down, is the best way to have the process handed off to the newest trainee in good shape. Written SOPs are learning aids and can be referenced frequently. An example of an SOP can be found in Appendix F.

Do not short-cut the duty of writing SOPs. This should not be asked of someone outside of the operation that the SOP is for. It is the employee's duty (the process owner's) to write down their processes and share with new employees. HR, or another function such as quality manager, may help facilitate the training, help with standardizing templates, and catalogue who has what skill, but they don't write the procedures. Simply stated, those that own the process should write their procedures down and share them with new trainees. Passing the buck on this only creates a cycle of non-ownership. Quality and possible safety problems are almost guaranteed in an operation where ownership of the SOP process is an issue.

Once the process training is bundled to skill sets and learning aids are developed, the HR or quality person in charge of training can build into every person's training plan the most common training needs such as orientation, quality, safety, asset care, and overall culture immersion. The training manager should also keep detailed records of who was trained and in what subjects for regulatory and disciplinary purposes.

Here are some examples of when quality is impacted if training is poorly planned:
- During orientation, the trainer forgot to mention the location of bathrooms and the importance of hand washing. This resulted in the brewery getting a cautionary note from the health inspector when the inspector casually asked a new employee where they wash their hands.
- Process training for malt milling didn't include inspection of the grind, and the quality concerns that come with poorly milled malt. A new employee opened the mill roller gaps to increase runoff speed, and results were unexpected extract and alcohol in the final beer.
- The HR manager forgot to mention the reward system for doing a good job in quality during orientation. A new employee was so embarrassed about their measurement that seemed high in dissolved oxygen that they didn't speak up, resulting in oxidized tasting beer for the entire run.

Finally, training and development plans should include leadership skills for all managers. Managers who have well-developed leadership skills form an excellent support network for a successful quality system. These managers are more likely to blend clear direction, coaching, and, when needed, encouragement. Managers lacking these skills, who don't recognize the need to grow and develop and empower their team, will see beer quality suffer. Be sure to support any training program with a good leadership development program. Absence of support in this area will be discernable in quality issues and employee morale, and staff retention will suffer.

Specific Training for Quality Managers in a Brewery

Basic
- Certified Quality Manager (ASQ)
- Certified HACCP
- Safe Quality Foods (SQF)

Intermediate
- Lean Management
- Six Sigma Green Belt

Advanced
- ISO-9000 quality systems
- Six Sigma Black Belt

Other HR policies that can impact a quality program include having clarity of roles, responsibilities, and accountability. When a brewery is beginning to grow, it may feel constrained to tie an employee into a job description. Most employees of small companies enjoy wearing multiple hats and getting a broad job done. It is convenient to be a little loose on the job description. However, when it comes to product quality, quality testing, and hold/release responsibilities, not having clear job duties creates conditions ripe for failures. A good example of this may be in the warehouse, where shipping and receiving duties fall on a select team or individual. They may be the last party to see product go out the door. If product is pulled from a cooler and the packaging shows signs of condensation, for example, the person loading the truck should know it is their *responsibility* to pull aside any questionable product. If not, complaints might come from the distributor who receives it.

When employees are educated on their roles and responsibilities in making quality decisions, then managers can hold them accountable. When employees are not informed and empowered, it is a struggle for the brewery to maintain control of the quality output. For example, a quality technician notified production that a beer held in the bright beer tank failed the bitterness specification. Unfortunately, the production manager was pressured to get the beer packaged. They didn't have time to wait for quality validation tests and began bottling the beer. It was determined the beer was accidently blended with a beer with higher bitterness levels at the filter, and the brewery had to halt bottling in the middle of the run and dump the bottled beer. The rest of the beer was blended down to the correct bitterness level. There were two failures in this case: The quality technician was never told it was their responsibility to place the beer on hold until they fully investigated the out of specification result, and the production manager failed to follow protocol by allowing the quality team to validate their results. This was a costly error, and one that could have been prevented with the proper training of both individuals at the start.

HR also helps set the policy of balanced accountability and discipline/reward systems. Most employees really want to do a good job. Making decisions for product quality may be in conflict if the message from management is to speed up production. HR can help ensure company objectives are clear, and that values are not in conflict. Having a clear disciplinary policy in place helps managers both when they have to hold employees accountable and when it is time to hand out rewards. If discipline is only handed out for failing to do a quality check, but rewards are not handed out, employees don't usually respond well. Employees need to understand that the expectation of their job is a balance of quality and efficiency. Employees also need to know that there is discipline for those who don't comply with expectations, and rewards for those who do. Adrian Gostick's *The Carrot Principle* is a great resource for managers on how to reward and discipline effectively.

Lastly, sometimes HR is responsible for goal setting. This may be generic across all the departments for specific company-oriented goals, such as new product launches, new engineering projects, or even output- and efficiency-derived goals. Many times a generic safety and quality goal is set for the entire operation. It may not be obvious, but the supporting functions of maintenance, HR, and even engineering all need to share in the overall quality goals for the brewery. As mentioned in the opening of this chapter, these functions do impact quality and have a supporting role to play in their policies and procedures, or lack thereof. Reinforcing quality routinely for all employees helps reinforce that everyone can impact the product in a positive manner.

ASSET CARE/MAINTENANCE

One of the most overlooked support systems that must be in place to ensure excellence in the brewery is asset care or maintenance management. The equipment used to brew and package directly impacts beer quality, and most well-trained brewers understand this inherently. Brewing schools teach many quality issues in which the root cause is poorly maintained equipment, high dissolved oxygen in packaged product, microbiology and spoilage issues, yeast health and fermentation problems, and packaging quality and seal integrity, to name a few. Even though a trained brewer is aware of the issues that will come about from poorly maintained equipment, they may not understand how to manage the program to keep their assets in top running condition. The focus here is to illustrate how a proactive maintenance program is supportive of quality and how to avoid a reactive maintenance program.

Every valve, pump seal, weld, shielding system, gassing system, pneumatic system, and moving part that performs poorly has the potential to make a lot of bad quality beer. "Reactive maintenance" is also known as "run until it breaks." New breweries may have to be reactive, due to lack of personnel or lack of planning downtime to properly care for equipment. However, this is a recipe for poor quality beer. Preventive maintenance is a better value to uphold. Appendix G shows the most likely quality failures from equipment maintenance. This list illustrates how the philosophy of "run until it breaks" has the potential to cause poor quality that goes undetected. As a piece of equipment is failing, the possibility of damaging beer over a long period of time is very likely. Therefore, "run until it breaks" becomes a sure way of making poor quality beer. If you are in the business of making beer, then you are in the business of maintaining your equipment (assets) that the beer is made on. The quality checks that work in concert with proactive maintenance are also mentioned in Appendix G. Using both quality check data and other maintenance data, such as vibration analysis, will put the brewery in a better position to not impact beer quality due to poorly maintained equipment.

The question becomes, "How do you prioritize, spend your resources most wisely, and choose what to proactively maintain?" This can be difficult to answer. One way to structure maintenance-focused priorities is to conduct an FMEA for the major pieces of brewery equipment. As discussed in Chapter Three, because an FMEA is a quantitative risk assessment, it is a great tool to help point the maintenance team to what equipment poses the biggest risks to quality if not maintained. The FMEA will show which failures have a low likelihood of detection, and which have the most severe impact on quality. Those potential failures become the focus for preventive resources, and either a more proactive maintenance program or better detection of the failure. Some proactive or predictive maintenance tests that world-class breweries conduct are acoustic or vibration analysis on pumps, oil analysis on gearing and motors, and lube rate of consumption.

To establish proper proactive maintenance programs, proactive maintenance requires:
- Dedicated on-site personnel that are trained in mechanics, pump function, and liquid transfer. They are also trained in brewing principles, empowered to conduct quality checks, and have a knowledge improvement plan to continually learn higher-level maintenance processes.
- A maintenance plan that details every piece of equipment and the repair work needed with a work order system. Preferably, this is on an annual calendar that corresponds with shutdowns for every piece of equipment.
- A well-organized spare parts "store."
- New equipment installation protocol and policy that involves the quality department.

Using the HR principles described earlier, maintenance team members should be included in training plan development. Not only do they have to understand equipment specifics, their understanding of brewing and the impact of the equipment's upkeep is key. Scheduling a specific training regime that involves maintenance personnel spending time in the brewing process and the brewers spending time with maintenance personnel is a best practice. Not only do they appreciate each other's work more and better understand how they can help each other, they are also able to communicate more effectively when there is an issue.

Maintenance personnel should be part of all quality team operations meetings and communications. Their workload, scheduled maintenance activities, and other priority work must be integrated into communications with the quality and safety staff. They need to know when to consult with a quality team member or a brewery operations manager with any questions about food safety or quality issues. This means they need the personality to integrate with most personnel and take in the entire scope of the operation. If maintenance is completely outsourced, as is the case with smaller operations, then it is the job of the brewery manager to be the liaison and communicate fully with the outsourced team. Defensiveness, second-guessing, or insubordination need to be addressed by upper management with any personnel in the brewery, but especially maintenance staff, as they impact quality, safety, and efficiency in the brewery. Some prime areas where communication gaps in maintenance practices can impact quality are: lube and chemical selection, major construction projects, and minor/temporary repairs. Some specific examples of these three areas, and why they are critical to brewing quality, are listed in the sidebar.

The other consideration to asset care and quality is the installation of new equipment. Once the equipment is installed, there is a period of time that the brewery will operate under "pre-control." This is a good time to dial in the new asset and its proper operation. A good rule of thumb, prior to purchase, is to budget 10% of the asset cost for monitoring the quality of the beer output from the equipment, either in-line or offline. Sometimes this monitoring isn't included in the equipment. For example, a centrifuge without dissolved oxygen monitoring is a great way to make bad beer. Dissolved oxygen can easily ingress into centrifuged beer from bad seals or line leaks, or from improper pressure differentials during the operation, causing pick-up at the bowl. This seems so foundational, but if the quality in-line testing equipment was not offered with the equipment itself, it is easily missed in budgeting. When new equipment is being ordered or designed, staff responsible for quality monitoring need to be involved in the planning, monitoring plan, pre-control phase plan, and engineering design.

Why Communication with Maintenance Is Crucial

Without proper training in brewing and food safety, many issues can manifest because of an employee that is uninformed about maintenance. Here are a few examples:

- In the daily rounds of oiling equipment, a maintenance employee picked up the wrong lube for the filler. Fortunately, someone in quality stopped them prior to using the lube. Using a non-food-grade lube in an area of the brewery that requires food-grade would have potentially contaminated product and caused product pullback. To prevent recurrence, the brewery made sure a quality staff member signed off on all new lubes and chemicals brought into the brewery, the maintenance manager made sure lubes were well-labeled, and maintenance employees were trained to know where the food-grade lubes must be used.
- Major construction projects such as cutting floors, painting, or welding cause dust, fumes, and debris to be kicked up into the air. These areas MUST be totally contained from the rest of the brewery. If it means dropping a tarp from the ceiling to the floor, do it. In one brewery, despite the quality team member warning, a maintenance manager began to cut out bad concrete without the level of contamination prevention needed. The situation was elevated to management and containment was fixed, but by then the damage was already done. This resulted in discarding an entire 200 BBL of beer because of contamination by wild yeast and bacteria. To prevent recurrence, quarantine areas must now be signed off on by operations and quality, and inspected by operations during the procedures.
- To properly clean and sanitize equipment, the equipment must be left pristine after maintenance. Not only should welds be clean and tool sanitation be expected, but maintenance personnel should be expected to keep track of their parts inventory during the job. In another brewery, after several months of investigating the root cause of microbiological issues in a tank, the operations team pulled the spray ball from the tank and inspected it. Inside the spray ball was a small drill bit grinding stone, partially plugging the spray ball. It was determined the drill bit was left behind after the clean-in-place (CIP) tanks were recently worked on, and it was flushed through the pipe to the spray ball. This one small bit caused many hours and months of diagnosing, testing, and even dumping beer. To prevent this from occurring in the future, the maintenance team was educated on the importance of inventorying tools and parts before and after each job. The manager took it upon himself to create a quick-check form for inventorying critical small parts during jobs, and reviewed these after each job was completed. If a part went missing, there was an extensive search in critical process equipment prior to the equipment going online.

A good example of integrating quality into the design of equipment can be as simple as consulting someone who cleans the equipment that is difficult to remove or clean around. For example, a brewery wanted to install a photo eye on a filler for missing filling valves. This eye was installed on a bracket that was integral to where cleaning had to take place. It was a great quality tool, as it was designed to stop the filler immediately if a filling stem came loose and fell out. (Because this was a high-speed filler, a failure from a missing stem could cause large product holds once discovered.) The engineers consulted with the brewers that had to clean the room routinely and came up with a quick-change arm that held the photo eye. This caused fewer issues at clean-up, and less downtime, but it also allowed for what could have been a harborage point for bacteria not to occur.

A good rule of thumb, prior to purchase, is to budget 10% of the asset cost for monitoring the quality of the beer output from the equipment, either in-line or offline.

Quality monitoring on new installations also can benefit operational efficiency, but this may not be obvious until a mistake is made. Putting an in-line meter for turbidity into the outlet of the centrifuge is considered so important that it is part of the centrifuge control. Still on a limited budget, it may not seem needed. A simple rule is if the cost of dumping a tank of beer exceeds the cost of a meter or monitoring device, go ahead and add the device. The likelihood that a tank will be dumped, or product lost, due to lack of monitoring should be taken into consideration, and it is usually a simple rule to follow.

The return on investment (ROI) may not be immediately obvious, or it may be hard to quantify for these types of project additions, so senior leaders need to maintain a hard-line policy that additions that help ensure a quality product get a budget with every new equipment installation. On a strict ROI, or cost avoidance scale, it is very easy to cut corners and wait for the failure and a major loss before the accountants are convinced it is worth the investment. Avoid this way of thinking and consider prevention your very best medicine.

SANITATION AND GOOD MANUFACTURING PRACTICES

Sanitation processes and supporting criteria of good manufacturing practices (GMPs) are a foundational requirement of every food operation in the United States. GMPs are a set of criteria that were developed to support the Food and Drug Administration's (FDA) code of federal regulation, 21 CFR. This code requires every food and beverage brewery to prevent the production of "adulterated" food via sanitation and general policy. This is a broad sweeping statement, but it means the condition of the brewery must be such that you have little likelihood of contaminating food or beverages with "defects."

GMP policies help prevent conditions that would support microbial growth in the brewery and have an impact on beer quality. The GMP criteria spell out specifics of what has to be maintained to control pests such as rats or mice, and the condition of the buildings and grounds. Therefore, general GMP compliance should be in the hands of a maintenance team member. If outside education is needed, certified sanitarians can be consulted or additional team education can be provided. Once a policy is in place, staff can keep a schedule and perform required tasks.

A simple rule is if the cost of dumping a tank of beer exceeds the cost of a meter or monitoring device, go ahead and add the device.

The GMP policy can also cover food safety requirements, such as keeping chemicals and lab glassware contained so they don't impact beer quality. A well-implemented GMP policy is very supportive of the quality mission and required for excellence in a brewery. HACCP (mentioned earlier) is a risk assessment and plan that ensures food is safe. GMP policy is usually a prerequisite for a successful HACCP program. General GMP policies for food plants can be found on the internet, and there are many excellent templates available. Many times, asking a trusted supplier of malt or other raw materials to share their GMP policy can provide great insight to improve a brewery's GMP policy. Appendix H contains an example of a GMP

Good Manufacturing Plan (GMP) Components

The typical GMP policy will have several key parts. Breweries should cover the following:

Responsibilities: State who is managing the GMP policy—including training and auditing.

GMP requirements: Remark on any internal standard, state, or what United States Code of Federal Regulations (US CFR) is cited. Basic GMP Requirements are provided in Title 21, Part 110.

Personnel: Describe how operations personnel will be managed to comply with GMPs, including: disease control, cleanliness, dress code, jewelry wearing, hand washing, gloves, hair or beard nets, storage location of personal articles of food and clothing, and where employees are to eat and take breaks. This can be difficult to discern in the brewery. Base your decisions on potential risk and document the basis for decisions. A good practice, but one not always enforced at craft breweries, is containment of food, hair, and jewelry where there are open vessels or packaging, such as bottles or empty cans.

Brewery and grounds: Mention how the grounds are maintained for pest control. Too many breweries keep doors open to the elements; unless it is a totally contained system, this shouldn't be allowed. Include any policies for door propping, open windows, etc.

Sanitary operations: Include general building maintenance policies, storage of cleaning chemicals, pest control management, and general cleaning procedures.

Sanitary facilities and controls: Describe policy for water supply, plumbing, and floor drainage that prevents sewage from getting into the facility. Also include directions for hand washing stations and rubbish disposal.

Equipment and utensils: State general policy that all brewery equipment and utensils shall be designed and of such material and workmanship as to be adequately cleanable and properly maintained, including anything made from wood.

Glass control: Include general policy for how glass breakage will be managed. This could also include lighting, bottles, and gauge glass.

Visitors and contractors: Describe how visitors are managed. Visitors and contractors must observe basic GMPs to ensure the safety of products and people. Anyone entering production or processing areas shall be asked to carefully read the Visitors and Contractors GMP Guidance.

GMP inspections: Describe how the GMP program will be validated.

Training: Outline how and when training is conducted.

Revision history: Include this as good document management.

The Master Brewers Association of the Americas (MBAA) published a GMP general template for brewers in 2013. It can be found on their website: http://www.mbaa.com/brewresources/foodsafety/haccp/Pages/documents.aspx.

program most breweries can use when getting started. (A more involved GMP guide can be found on the Master Brewers of America's website http://www.mbaa.com/brewresources/foodsafety/haccp/Pages/documents.aspx.)

GMP Culture and Implementation
Writing a GMP policy is only a small portion of the program. The implementation of GMP policy is when the rubber hits the road. There are many strategies to implement a successful GMP program. If you are

just starting a new brewery, having a GMP policy in place before the first day greatly reduces the difficulty of implementation. All employees read and comply with the policy as part of training before they start work. If this opportunity has passed and you are playing catch-up, you will have more work to do, especially if you are operating with little to no written GMP protocol in place. A classic "plan, do, check, act" strategy will help implementation of a GMP for a large communication rollout, a period of employee adjustment, check after the adjustment period is over, and HR coordination of discipline or rewards when the brewery complies with the new policy.

Many larger breweries provide teams with incentives for complying with new policy. Making a game out of it can be a fun way to integrate GMPs. Becoming too punitive too soon will create backlash. The goal of implementation should be to make employees feel good about changing bad habits and instituting good ones. Taking jewelry off and not chewing gum or drinking soda in areas where it is not allowed takes time and practice to create compliance. Use your imagination; make it fun and safe for the employees to learn.

Senior Leadership must also have a clear understanding of the GMP policy and lead by example. As part of the Quality Manual, senior leaders should sign off on the GMP policy. This means they cannot chew gum or wear jewelry or open-toe shoes where it is not allowed according to the GMP policy. Setting a budget aside to support GMPs as part of maintenance is another way brewery leadership can visibly support the policy. As a brewery ages, there are usually many areas to improve related to GMPs. Any time an area becomes easier to clean, more organized, or less cluttered, this will save time and money, and improve safety. Most senior leaders should need little convincing of this. Leaders can also consider bringing in an outside GMP expert to routinely audit and encourage the production staff to continually improve. Audits can result in large to-do lists and senior leaders can be part of the process by working with operations management to prioritize the list every year and cover what the brewery can realistically accomplish.

Sanitation Planning

Sanitation procedures are required to maintain clean and sanitary conditions. These procedures include cleaning everything—floors, drains, walls, tanks, fillers, fans, overhead beams, and even forklifts. Every sanitation procedure should have four basic components:

1. Chemical types for each situation and the corresponding safety information
2. Outline of the procedure, who conducts it and how often, and frequency (time, temperature, and chemical concentration)
3. Validation step that the cleaning was conducted properly
4. Documentation of sanitation

When one stops to think about every nook and cranny to be cleaned in a brewery, it can be mind-boggling to write a procedure for everything. To help organize the chaos, a Master Sanitation Schedule (MSS) is usually created as part of planning sanitation. This is one complete list of everything in the brewery that must be cleaned, by whom, and how frequently. Once the MSS is written, procedures can be written. The MSS and procedure writing is an ongoing process. It can be a massive undertaking, so some sort of prioritization is recommended.

Start with the most risky areas of the brewery and move to the least risky. A procedure for cleaning the stairs should not take precedence over cleaning the mill room, for example. Writing the MSS list and the procedures should be mostly the responsibility of the brewery maintenance team, not the quality team or the operations team. This will seem controversial to many, but the maintenance department ultimately has to deal with improperly cleaned equipment when it breaks down. A maintenance employee, for example, does not want to work through layers of mold to access key areas of a conveyer, which can and will happen if it is not maintained properly. Also, maintenance is typically responsible for the condition of the buildings and grounds that surround the brewery, so it makes sense that they stay involved at this level of management.

The maintenance manager isn't alone in this endeavor to capture the sanitation plan on one MSS. An MSS takes a joint effort to pull together. Input provided by quality or operations departments may include noting when and where mold buildup occurs, what sanitizers are safe to use on certain equipment, and what the policy is for sanitizer levels. Input from operations may include scheduling and determining who is skilled in the cleaning. Sometimes it is most cost efficient to bring in an outside contractor to do annual or bi-annual cleaning.

Tank cleaning or clean-in-place (CIP) procedures are conducted so frequently that they will be the responsibility of the operations team. CIP procedures deserve their own special attention to ensure they are written correctly. Your chemical supplier can help you do this. Be sure to include cleaning chemical types, temperatures, flow rate, and sheeting activity of the chemical on the tank, water chemistry, and even spray ball type and maintenance. Lastly, be sure to include sanitation processes in training.

Validation
Validation is a critical aspect of sanitation and shouldn't be ignored. The quality manager can help educate the operations team on how to validate that the cleaning was conducted properly, but it is not recommended that this be made a separate functional role. Instead, have the same person who conducted the cleaning validate that it was done properly. This is similar to performing a quality check on a process. For example, the person who cleans a tank should be responsible for validating that it was cleaned by using a microbiology swab or an adenosine triphosphate (ATP) swab quick-check. (ATP swabs are a luminescent test to check for general cleanliness and a good quick-check in place of microbiological testing of every surface. Check out ASBC Method Microbiological Control 6.)

In addition to quick-check tests using swabs (either for microbiology plating or with ATP test swabs), visual assessment is another way to validate if a sanitation procedure was conducted properly. If using any of these methods be sure to include them in the process training. It can be easy to misinterpret exactly what should be looked at critically for a visual assessment. When cleaning a conveyer, for example, it is common to find shadow spots, or areas that just don't get well cleaned. These are good places to use in verifying completion of a good cleaning job. Include any "hot spots" in training materials.

RECORD KEEPING
Record keeping can become burdensome if not properly handled. A brewery keeps many forms of records:
- Policy and procedure
- Templates for data records
- Training records
- Inventory records
- Maintenance records
- Quality checks and corrective action records
- Consumer complaints

Keeping templates and completed records organized requires some oversight. Having a point person that maintains templates, how revisions are logged, where records are kept, and for how long, is essential. Because this can be a comprehensive duty that crosses departments, such as maintenance and HR, having one template for SOPs, policy, and records is extremely helpful as a brewery grows. An example of a template is included in the appendices as the header for each example.

Complaint Records
The last component providing support to a quality program is to have a way to record and monitor complaints. The quality department may fully manage this area, or in small operations it may be sales and marketing that funnels the information initially from accounts or distributors, or even consumers. Regardless of who is on point, the data must be recorded on a form and communicated to operations in a timely way. The operations and quality departments should routinely review summarized data to get to the root cause of problems and address concerns. Senior leaders need to review the data and ensure their quality philosophy and direction aren't at odds with the customer's changing needs.

Setting a quality goal on complaints is a great way to keep track of whether the brewery is underperforming or successfully meeting expectations. Usually complaints are logged by computing a ratio or index to the amount of beer produced. For example, the number of complaints in one year for every 10K barrels produced (<10/10K BBL). If the brewery is smaller than 10K barrels, it may not be receiving complaints in an organized manner. Benchmarks for the industry on this metric are not well published; however, as a brewery grows its complaints need to come down. Once the brewery exceeds 10,000 barrels, it needs to organize who receives the calls, and who compiles the data. Someone should be designated the point person for fielding and logging all complaints as they come in. Megabreweries may outsource the handling of phone calls but maintain a sophisticated tracking of consumer complaints by lot code. The downside of this approach is that the call taker may inaccurately identify the complaint and code it incorrectly, resulting in time wasted to investigate a nonexistent issue. Still, keeping track of complaints is highly advised and it will help drive overall quality improvements in the brewery.

The brewery is a varied place in terms of work functions. Despite this, all work functions have a role to play in improving overall beer quality. Keeping track of consumer complaints can maintain the focus on quality.

KEY TAKEAWAYS

1000–15,000 BBL Brewery	15,000–150,000 BBL Brewery	150,000+ BBL Brewery
• Supporting functions are just being defined, some may be outsourced. Be sure these are included in communications about quality.	• Supporting functions are new and may not be broadly integrated into the brewery through all departments. As processes are formalized, be sure the link to quality isn't lost.	• Supporting functions now have formal processes. • Routinely review and critique goals related to quality.

FIVE

STRATEGIC COMPONENTS IN THE QUALITY PROGRAM

In the last chapter we discussed the policies and procedures that support daily operational needs and that assist in the success of the quality program. Through these supporting elements, departments such as Human Resources and Maintenance have accountability toward overall beer quality. This chapter's focus is also on supporting components, but they are more strategic in nature and are likely to be facilitated by the quality department in the brewery or by a corporate quality function in a larger organization. Specifically, we will cover the introduction of new products and the implementation of structured problem solving to assist in troubleshooting.

NEW PRODUCT DESIGN AND IMPLEMENTATION

Back in the mid-1980s and early '90s new products were somewhat of a novelty in the brewing industry. Flavored malt beverages and new packaging configurations made a splash, but the major players were focused on outselling one another, consumed in marketing battles, not necessarily battles of innovation. Still, some innovations helped boost sales even for the larger brewers. Adding a flavoring to a lager became a new way to infuse unique flavors, such as lime peel, into beer. Additionally, there were small breweries beginning to challenge basic assumptions about beer flavor by using old processes with a new flair. Processes such as aging in oak barrels, utilizing ale yeasts, adding herbs and spices directly into the kettle, and even dry hopping were all older techniques experiencing revitalization by a wave of brewers interested in broadening the flavor spectrum of beer. Fast-forward two decades and this small group of brewers has influenced the entire category.

"SKU proliferation," or the adding of new beers or packaging configurations to a company's offerings, has become the norm. With so many new product and packaging introductions, it is hard to keep up on the innovations. Aging in barrels brought with it the risk of exposing other beers to what were once considered spoilage organisms—*Brettanomyces*, *Lactobacillus*, and *Pediococcus*, for example. Dry hopping became wet hopping, and hops began being used in ways never before considered. Of course, with innovations come new unknown risks to quality. Innovation is occurring at such a fast rate, it can be a struggle to ensure no quality issues arise from tests or trials. This can result in pulling back spoiled product from the shelf. Additionally, with new ingredients rarely used before in brewing, food safety controls are challenged. Pulling the reins of quality and food safety risks slows down the pace of innovation. Product innovation is the one area where speed can be directly at odds with quality and food safety. To manage this natural conflict, good design and risk management will help create higher quality and longer-lasting innovations.

To achieve the right balance of speed to market and quality, innovations should be grouped into four common categories, after which design and risk assessment processes will follow. Product innovation is typically grouped in categories as follows:

Incremental: The incremental innovations are market focused.

- Incremental improvements – These are core capabilities to the brewery and improve on quality or consistency in the beer. Examples may include utilizing a different type of hop, or changing a malt input to extend shelf life. In packaging, this may be putting a product that was once only bottled into a can (assuming the can line is already in the brewery).
- Incremental changes – These are extending product lines or markets. For example, this may mean a brewery innovates on a new way of shipping across the country.

Radical: The radical innovations are technology focused.

- Radical improvements – These are also core-capability focused, but extend the use of new technology into the process. Examples include adding new hop types or malts into a new beer for the brewery, or using a new delivery vehicle (hop extract vs. hop pellets, for example) to add ingredients into a standing beer. In packaging, this may be a new type of package design.
- Radical changes – These are the disruptive new products that change our understanding of beer. It may include new technology; it may be bringing back old styles, or re-inventing styles. The craft beer industry is most known for bringing radical change type innovation to the brewing world. A radical change was using wet hops in beer, for example. (Loch, 2008, 72)

Note that improvements are core-capability focused, while the changes are opportunity focused. One can look at improvements as the projects that save the company money, and the changes as the projects that bring in new opportunities of revenue. Both support a better bottom line. The key is to be sure you have identified the time and resources that need to be allocated to each type of innovation and that you have determined the mix of innovations your brewery can take on. This is the basis of an innovation strategy. For example, a brewery that likes to bring in radical changes may indeed have 80% of their innovation projects in new products. A brewery that is doing this much product innovation should be suited to innovate and trial new products through a brewpub or small-scale customer testing. Radical changes and improvements need to be given the proper time and test channels to fully vet any unknown risks. As projects take shape and products are designed, those responsible for marketing the new product need to be sensitive to test marketing—as not only a means to prove a concept, but also a way to ensure that the quality of the product remains intact. A person dedicated to quality should be involved at the design stage to communicate the length of time a risky project will need to be in testing.

Additionally, though the quality department may not define the innovation strategy, it needs to have a keen understanding of it so that it may enforce it. Too many radical change innovations introduces risks to existing products. Therefore, quality is indeed a large stakeholder in the strategy. It should be discussed and agreed upon by involved departments so as to be clear on how many radical change projects a brewery can successfully take on at one time. Radical change innovations are not the only issue. Simply having too many innovation projects (including improvement projects) coinciding can also impact product quality. As innovations become layered, even if incremental, the changes to standard process can become difficult to communicate and end up being very disruptive to an operation. For example, a brewery decides to make a new product with spices as ingredients. For safety reasons, the spices will be loaded into the kettle via an automated injection. This requires engineering, design, plans, and resources from the brewery to help coordinate. If the brewery decides to also begin another innovation in the same area of the brewery, say, an incremental change to the kettle design, the personnel dedicated to the first project will not achieve their goal in a streamlined and efficient manner. Therefore, be smart about taking on new innovations. Identify how many incremental/radical changes or improvements you need to make in a calendar year, and then allow the appropriate time for each project. Optimize the flavor or the equipment before committing to repeating the process in many products. (See Figure 5.1.)

Managing Innovation and Resources
New innovations, even if they are not opportunity focused, take time and attention from staff to ensure other products are not at risk for loss of quality control. One key factor for success is to be sure brewery leaders have a strategy and understanding of the operational constraints in order to pace new innovations. Once the strategy is decided, the operations team can act. Risk review and planning are necessary for individual projects and building the synergies of consecutive projects. Sometimes one innovation can be a big commitment, such as adding in new capabilities in the brewhouse. Add two or three at once, and failure of some sort is more likely to occur. Therefore, as part of the process, it is important to take a big-picture approach to all the projects, new launches, and innovations being managed at one time.

As projects take shape and products are designed, those responsible for marketing the new product need to be sensitive to test marketing—as not only a means to prove a concept, but also a way to ensure that the quality of the product remains intact.

The quality team may be asked to organize a broad view of innovations by using a calendar or other type of visual to communicate the big picture of the job at hand. (See Figure 5.1.) Operations managers can have their team members weigh in on "what can go wrong" if multiple innovations were to happen at one time. Once timing is properly determined to help reduce risk in overlapping projects, you can assess the risks of the individual projects.

The best tool for assessing an individual project is the FMEA risk assessment mentioned previously. Because FMEA can be time intensive, this level of risk assessment may be used to assess "radical change" types of innovations in the brewery. For example, if the brewery typically only makes lager, and a new product is

	Jan	Feb	Mar	Apr	May	Jun	Jul	Aug	Sep	Oct	Nov	Dec	Jan
Product G Can graphics				■	■	■	■						
Product H New label					■	■	■	■	■	■			
Product K New label			■	■	■	■							
Product L New product		▨	▨	▨	▨	▨	▨	▨	▨	▨	▨	▨	▨

Figure 5.1: Example of a Calendar of Innovations Strategy   Legend: ▨ New products 25%   ■ Core capability 75%

designed to combine lager yeast and another yeast, this is a radical enough process change to look at a formal risk assessment to help capture all the what-ifs and subsequent testing and controls from a quality perspective.

It's also important to remember that the pace of new introductions or new innovative techniques is not only market driven, but should also be balanced with the operations team's ability to bring on new products. As an operation becomes more adept at bringing on new products, they'll be able to bring in more projects at a faster pace. The brewing team's ability to analyze risks improves with experience, as well. Therefore, if the brewery is not typically asked to perform risk assessments, adding too many to their plate too soon is a recipe for quality disaster.

New Product Introduction
The quality team has a special assignment when a new product or packaging technology is brought into the brewery. Assuming that the design and risk assessment had quality personnel input and the project is going forward, the next item to fall solely on the quality team is tactics. The following six are good general guidelines:
1. Lab preparation: Add new tests to the QC lab, calibrate the taste panel.
2. Add documentation to records: new ingredient specs, new finished-good specs, Kosher paperwork, MSDS and safety paperwork, certificate of analysis for alcohol, a beer description for taste panel.
3. Create the quality test plan for the first production.
4. Train and communicate all new specs, new testing protocols, and testing phase of the control plan.
5. Follow up on how the product is performing in the market.
6. Adjust any quality checks in frequency, as needed.

If the risk assessment (e.g., the FMEA) identified that a new product will require a quality test that is new to the brewery, the first challenge is to determine a timeline for when the test will be implemented. The timeline for a new product introduction that requires new tests will need to incorporate any new test equipment delivery, quality staff training, and a measurement optimization period. For example, if an unusual yeast produces certain flavors that are detected by the gas chromatograph, and it is assumed the quality team will test for these flavor markers during the initial production batches, then the quality team must be able to communicate their timeline to accurately bring this test online and make sure their instrumentation is validated. Mentioned previously is the period of pre-control with any new equipment delivery in the brewery. With new test equipment in the lab, plan for a period of adjustment to optimize the test, validate it, and train the lab technician accordingly. Don't assume this can happen during the initial productions of a new product; it will cause delays. Calibration of the taste panel is similar to lab preparation. A standard test sample of the new beer from a pilot brewery can help. The taste panel can become familiar with the new beer, describe the flavors and intensity, and create a true-to-type descriptive profile of the new product. Once a flavor profile has been agreed to, the panel is ready to provide the quality data feedback.

The next task is updating all records, templates, and documentation. This isn't the most exciting part of new product introduction, but it is an important step. If it is missed, the operation is left to speculate on new specifications. For example, if a brewery receives a new raw material, such as a flavor or herb that will be injected downstream of heating, a microbiological specification is important. (See Figure 5.2.) Sometimes the supplier's capabilities are standardized to different industries, and they are not ready to bring on the requirements for brewing. If not specified, the higher load of bacteria that the product normally contains can cause major headaches for a brewery if not specified in advance.

The quality team's third step is to create a quality control plan for the new product. This should be based on information gathered by the risk assessment. If the brewery has a corporate body overseeing the risk assessment, the brewery quality team should be asked to create the new test plan for the product. Only the brewery team members really understand the process and product risks, and how they may interplay. The

Microbiological Specifications for New Ingredients

Brewers are increasingly adding a multitude of new types of ingredients to beer as part of new product development. Assuming they are approved for food use, these ingredients may seemingly be safe and harmless to the brewing process. However, brewing microbiology is very specific and should be addressed in microbiological specifications of raw materials.

A microbiological specification of non-detectable total plate count or <1cfu/100 mL is commonly reported in foods. However, the tests to conduct the total plate count can be generic and not designed to test for the microorganisms that can be found in beer.

Therefore, check with your supplier on what types of bacteria, yeasts, and molds they test and track. Many times they need to be asked to test for lactic acid bacteria, yeast, or mold. Don't be shy—ask!

Figure 5.2: Example of a Certificate of Analysis (COA) for vanilla paste.

quality control test plan may be a simple chart that shows the change to the standard QC plan for the new product (and not the entire QC plan). The plan should indicate if it is a temporary plan during pre-control or a permanent change for the new product. This can be used as a means to communicate to the brewery what the new product is, and what tests are required. (An example of a control plan summary for a new product introduction is found in Appendix I).

This brings us to step four, which is all about communication and training. Using the control plan summary, provide a quick synopsis of the new product, what it looks like, and why it is different to help employees feel more comfortable with the process. Remember, they are the eyes and ears of quality. Providing them an opportunity to give input, ask questions, and engage in new products will bring great benefits. Don't shortcut communication if you want the launch to be successful.

Adjustments can always be made in the brewing process to dial in a new beer, but the consumer can be hard to win back if lost.

The final two steps the quality team should perform are monitoring the new product and gathering consumer feedback. Adjustments can always be made in the brewing process to dial in a new beer, but the consumer can be hard to win back if lost. Keeping an ear to the ground, watching over pre-control data, and making adjustments where needed will close the loop in a new product introduction and ensure success.

IMPLEMENTING A STRUCTURED PROBLEM-SOLVING PROGRAM

When a quality professional asks if a manufacturing company has a problem-solving program, they are not referring to the problems you have, or whether you address them individually. Instead, they want to know if the business has a formal tool to track, govern, and oversee that everybody in the organization is engaged in solving the business's problems. Brewers problem solve often. If a brew does not taste as it should, does not hit target gravity, alcohol, bitterness, or color parameters, there is a problem to solve. If these problems occur frequently, and begin to cost the brewery in re-work, quality holds, or loss of customers, then it is time to roll up the sleeves and look at formal problem solving as a solution to these common ills. Problem solving is a team activity that, when done properly, will not only eliminate these sources of extra cost and losses due to poor quality, but will help you find ways to save the business money, in areas that were not considered problematic. Imagine if you had a list of all the problem spots in the brewery to solve, and for each problem you had a team assigned with a well-trained leader who used facts, data, and analysis to get to the root cause of the problem. Finally, imagine if each team used its collective knowledge to brainstorm and find the optimal solution. That is what a good, structured problem-solving program looks like, and if a brewery is ready, it is a powerful strategic tool to have in the quality program.

A formal or structured problem-solving program is a critical element for a brewery if it wants to save critical resources, reduce costs, and improve quality.

A formal or structured problem-solving program is a critical element for a brewery if it wants to save critical resources, reduce costs, and improve quality. Formal problem solving isn't a requirement for growth, which many breweries focus on in the early years, but it can help reduce issues during growth phases. Unfortunately, because it isn't required for growth, problem solving is often overlooked in business strategy sessions. Structured problem solving is often ignored during strategy sessions because it seems to be hard work, for big companies only, or it is not important enough to take foothold against other demands of a brewery operation. Many times, a major quality issue that results in a costly product pullback is the trigger that pushes a brewery into formalized problem solving. So, when is it advised to implement a structured problem-solving program? If you find that the brewery continues to have problems that return and plague the operation, then that is as good a time as any to assign this duty, or bring in an advisor to assist. Usually after the 10,000 barrel mark, the brewery is

Problem-Solving Program	Benefits	Risks
5-Why	Simple, straightforward	Seems tedious for solving simple problems. Not practical for difficult problems.
Root Cause Analysis (RCA)	Good training and discipline to find root causes. Good for smaller breweries to learn about determining a root cause before spending money to fix a problem.	Doesn't provide a lot of structure, and can be documentation heavy.
PDCA (Plan, Do, Check, Act)	Simple, logical	Not a lot of structure for solving longer-term problems.
DMAIC (Define, Measure, Analyze, Improve, Control)	Well-structured, can be flexible for solving problems ranging from easy to hard. Most data-driven problems can be solved using the simple data plotting tools offered in DMAIC. Great for breweries that have lots of data but do not solve problems well.	Requires training and a means to communicate progress.
Six Sigma/Lean Manufacturing	Very well structured	More expensive for training. Can be overwhelming to initiate as a program. Need commitment beyond one year to complete.

Figure 5.3: Formal Problem-Solving Programs. This figure shows several types of formal problem-solving programs and the risks/benefits of implementing them.

making enough beer to run into costly problems, and has a large enough staff to make uncoordinated problem solving an issue. Therefore, it isn't advised to wait too long to bring some sort of formal problem solving into the operation that can be improved upon over the years, even if it is very basic.

A formal problem-solving program has many side benefits. For one, it is a means to communicate key issues of the operation, including quality problems, to upper management. It is also a great way to boost morale, and bring a new wave of employee engagement to the brewery. Furthermore, it helps prioritize the seemingly numerous issues that can bog down an operation, and it helps get more done with fewer resources. The company that focuses resources a mile deep and an inch wide versus an inch deep and mile wide will prevail in the grand scheme of things. A formal problem-solving program helps do exactly that.

So, how to get started on formalizing problem solving? First, choose a path. This will require some research on behalf of the leader instituting the program. Some may decide to bring in a consultant who can help determine what path is best. The quality manager or director in larger breweries should be a resource to tap to research this for the organization. Because this requires knowledge of the brewery, the culture, and the complexity, it is highly recommended that operations work with the team developing the program. The formal problem-solving programs most referred to today are listed in Figure 5.3 with the benefits and risks involved in implementing each one.

As Figure 5.3 shows, there are many advantages and possible disadvantages to instituting different programs. The best advice is not to bite off more than you can chew. Starting with the simple programs doesn't hurt, as the knowledge can evolve. Problem solving should be looked at as a journey. It takes years of drumming the same beat to get it right in a manufacturing environment. Pick a path and plan to stick with it for several years.

You can be assured that the paths are similar in many ways. Each of these programs follows the same formula to solve problems, though they differ in the level of training, expertise, and depth of problems they are able to solve. The formula for problem solving is a logical path to answer any question. The first steps are to define the problem clearly, analyze the data involved, and determine the root cause. Once the root cause is determined, the problem can be solved with one or many solutions. These solutions are communicated and implemented, and then followed up on to ensure they stick. These steps were summarized earlier in the most basic of problem-solving programs as Define, Measure, Analyze, Improve, Control (DMAIC).

Because this is so formulaic, no matter which path you take you can rest assured that if you change to a new program at a later date, the initial training will not go to waste. The common threads that hold a problem-solving program together are: teamwork, communication, continuous improvement, and rewards. These are the foundational values for any problem-solving program.

> **Path to Formal Problem Solving**
>
> A path that is recommended for a growing brewery to start with is 5-why or root cause analysis. Use a template and train teams to work together to document their problem-solving thought processes. The next step up is a DMAIC or PDCA program. This usually requires that one or more team members become trained in the tools to analyze data such as statistics. The pinnacle is implementing a Six Sigma problem-solving program. This is a very formal and sometimes costly but effective program that, once sown, can reap years of cost and quality savings for a brewery.

The next step in formalizing your problem solving is to train and communicate the new way of operating. This is a key step for breweries in particular. There can be many problem solvers in a technical facility, and it can become easy to simply lean on them to solve all the problems. But the downside of this is that they do not share their thinking, so the team feels left out and morale suffers tremendously. To avoid this pattern, train all employees to be problem solvers using the method you want to see employed. This may require an outside consultant to help with plan development and training, or in a large facility, to take it in a series of steps. Start with a pilot team, train a facilitator and key team members and give them a small project to try out the methodology, and then refine and expand. For resources on problem solving, see the Resources section.

Once the training and pilot team has kicked off, then it is time to communicate the change to the rest of the brewery. This is where a senior leader or brewery owner MUST be involved. For the change to stick, the expectation must come from the top. Sometimes top leaders need a nudge to see the strategic element to problem solving and make it a priority. Showing examples of leading companies in your area that utilize Six Sigma, DMAIC, or other formal problem solving is a good way to bring them to an understanding. The brewing industry tends to insulate itself from other manufacturers. Quality and problem-solving goals are a great chance to open the doors and learn from others. It saves money, improves quality, and boosts morale. The downside of not acting is poor quality—and, sometimes, increased chaos.

Problem solving that addresses an issue but not its root cause will cost the company money. By making it clear that one method will be used to solve problems, and to communicate the problems that have been solved, the brewery will benefit. Now all the issues on the list can be managed and prioritized efficiently. The last step is putting governance to manage this list in place.

> Quality and problem-solving goals are a great chance to open the doors and learn from others. It saves money, improves quality, and boosts morale. The downside of not acting is poor quality—and, sometimes, increased chaos.

Managing Problem Solving

Once the direction is set and training and piloting in place, before communicating, the last step of planning is to determine how the leadership of the brewery or organization will remain in touch with the progress of the teams and improvements they make. This is governance and communication. The most basic governance found when applying a problem-solving model such as DMAIC would have three levels: champion, sponsor, and facilitator. A senior leader can function as the *champion* of the projects by prioritizing problem solving and ensuring everything that is on the list is strategic and important to fix. The sponsor can be a manager or supervisor, taking direction from the champion, and *sponsoring* a team to solve the problem. The sponsor remains close to the project and is ultimately responsible for the output. This position takes responsibility for the team's decisions and timeliness to find the root cause and implement the fix. Lastly, a facilitator organizes the team, trains them, and keeps them on track. A team leader can function as *facilitator* of the team. In larger facilities, there could be an additional governance team that oversees problem solving and ensures the process and plans are being implemented. This team is not doing the problem solving per se, but operating as a coach or guidepost team for the brewery. It is an effective way to keep consistency. This team of sponsors can bring in the facilitators to review their projects, and learn from one another on a bi-weekly or monthly basis.

For example, imagine a situation where a small regional brewery has had multiple complaints about spotty irregular keg cleanliness. The quality team remarks on these complaints to senior leaders. Because keg sales are key to the business plan, the owner of the brewery asks for the problem to be solved. The supervisor of packaging would ask the keg team lead to put together a few other staff members and solve the problem within a certain time period. The team leader must be trained to facilitate problem solving, or ask a quality team member to facilitate the team through the problem-solving process that the brewery employs. The facilitator helps the team through the steps of their process: listing possible causes, determining what to measure, analyzing the data, and getting to the root cause.

In this case, after visually measuring the keg spears after washing cycle, the team discovered the float was being returned with some very old kegs from one distributor. These kegs did not get adequately washed in the wash cycle. The fix was to add a check upon receiving kegs from that distributor and cull out old kegs from the float and set them aside for special washing. After discussing their observations, root cause, and recommended action with their sponsor, the supervisor agrees to the root cause and solution. He or she would help implement the new check in receiving. After two months the team checks back with the quality manager and ensures their complaints have gone down to a manageable level. Once the project is completed, the champion brewery manager or another senior leader may provide a special lunch or gift to the team for their work.

This scenario made sure the key elements were in place: teamwork, communication, continuous improvement, and rewards. It was a team effort to look at the possible root causes, measure, analyze, and come to a data driven conclusion. Note that a single employee isn't credited with this fix, but instead, the entire team had full authority and responsibility to solve the problem together. The sponsor helped the team communicate their progress, and helped another department improve their processes. Lastly, the champion made sure they were rewarded. Keeping projects front and center of the brewery by having a problem-solving visual poster board or electronic sharing portal is another way to communicate and spread the culture of improvement.

We covered some strategic elements of the brewery that help support quality, namely introducing new products in a controlled way and establishing formal problem solving. Though establishing a good problem-solving system takes time, it is a layer on the quality program that can make a significant difference in the brewery, in quality outputs, and also in culture. It is worth the time, effort, and thought process to put a program in place.

KEY TAKEAWAYS

1000–15,000 BBL Brewery	15,000–150,000 BBL Brewery	150,000+ BBL Brewery
• New products are being introduced as the brewery grows and is learning how it fits in the market. Be sure quality criteria are being set, even for something "one-off." • Problems are beginning to need solving, so taking them one at a time as a team is a good approach. Use data as much as possible to help get to root cause.	• New products may need slower introduction and more consumer tests. • Quality should be at the forefront of any new processes being developed. Tests are going to need to be developed for new risks. • More formal problem solving may be started. • Start building staff skills for problem solving.	• New products should have a formal review process. • Problem solving may become a part of the quality goals and culture.

SIX

THE BEST TESTS FOR A BREWERY

This chapter will cover the basic tests most breweries must perform. Though the information in this chapter may be rudimentary for some readers, the basic QC tests are the lifeline of the brewery and can make or break how well a lab is run. These tests require deep fundamental knowledge of how they work and why certain procedures are done when conducting the tests. Most people that have worked in a lab will know there are nuances in technique and process in even the most simple of tests. While technique and skill can be learned by trial and error, good management will ensure the right technique is passed on to new employees through training (and that it doesn't have to be continually re-learned). The brewery may continue to operate if these tests are not done, or not done well, but at some point large deviations from the normal process will not be found until it is too late. Therefore, because the results from these basic QC tests result in management decisions, and process adjustments, it is worth the time to dig into the basics to assure the test results are reliable and credible.

The number-one goal of any laboratory manager is to produce accurate and precise data. That way, even if the data shows trends are not good, it is acted on. Lack of action is a problem, but even worse is having bad data, leading to bad decisions and increased costs due to wasted efforts or loss. Another goal for a brewery laboratory is to ensure it is adding value with every test. Conducting too many repetitions or too many non-value-added tests, such as checks that duplicate each other for no reason, or testing for a variable (e.g., a certain microorganism that once was a problem, but no longer is) reduces the credibility of the quality program. As mentioned previously, the QC plan has to be occasionally reviewed for test efficiency, effectiveness, and to determine if it is worth investing in new technology.

An entire book could be written on brewery QC tests, so let's approach this from a management lens. Namely, answering the questions, "What is critical to know when managing people in the laboratory?" and, "How do we ensure test accuracy to help prevent every QC check being questioned and re-tested?" We will dig into the nuances that make the test reliable or not, and discuss quality assurance best practices. We will also briefly discuss what skills are necessary to perform each test. (Appendix B illustrates where each test is found as part of a QC plan.)

The Brewery Laboratory
The brewing lab requires some specific services as it grows. Therefore, if one is being designed or considered in a new brewery, be certain to plan for these variables:
- Water – City water is sufficient. Design the plumbing to code. Adding a coarse water filter on the water supply to the lab for later installation of a water purifier isn't a bad idea, either.
- Water purification – If the brewery is small, purchase distilled or de-ionized water for all tests. Eventually a system may be installed in-line.
- Counter space – Like a good kitchen, loads of counter space can feel like a luxury—that is, until there is so much equipment it is no longer enough. If starting small, be sure to overcompensate on counter space.
- Power – Ensure that there are adequate outlets along the laboratory bench, along with the ability to add outlets (up to 240 V) as the laboratory grows.
- General space requirements – A lab can fit in a small area, assuming it is heated and cooled sufficiently and has access to water. Space to eventually store gas cylinders is ideal. Air and gas (CO_2 and natural gas) are nice to have for future expansions.
- Equipment – Besides testing equipment that will sit on top of the counter (which will be dependent on the brewery's budget), you can add a refrigerator (that maintains excellent temperature control), a dishwasher with stainless interior (lab-grade preferred), and a double basin sink. See sidebar for more detail on equipment.

> **Basic Equipment Setup for Breweries by Size**
>
> - **Start-up 0–5000 BBL** – Microscope (for yeast checks and sediment of bottled/canned beer), pH meter, pipettes, test tubes, hemocytometer, slides, cover slips, microscale (weigh to 0.01 g), and small water bath (60° C).
> - **Mid-size brewery equipment 5000–75,000 BBL** – Add an alcohol or density meter, spectrophotometer (BU, color, and many other simple checks), titration setup (calcium, titratable acidity), incubator, and sterilizer/autoclave (for microbiology), Laminar flow hood and chemical hood.
> - **Regional brewery equipment (75,000+ BBL)** – All items above plus at least a gas chromatograph (diacetyl, VDK, esters, etc.), CO_2 and dissolved oxygen testing equipment.

- Floors/walls/counter materials – Floors can be concrete or otherwise durable to chemicals. Counters may be made of linoleum; however Corian® or lab-grade counter surfaces may be found and used. Walls should be smooth, if possible.
- Hood – Required when working with volatile chemicals that can be a health risk. Laminar flow hood for microbiology analysis is required for specific testing. (A small brewery doing limited microbiology testing does not need this.)

MICROBIOLOGICAL TESTS

One of the most basic and critical tools in the brewery quality program is the microbiology testing program. Microbiology tests include identification of organisms with microscopy (use of the microscope), cell staining, hemocytometer use, plating on various mediums, and possibly the use of other quick checks, such as adenosine tri-phosphate (ATP) swabs, and even polymerase chain reaction (PCR). These tests are addressed below.

Microscopy

The microscope is arguably one of the most critical tools in a brewery quality lab. It is used to check yeast health and viability as well as slurry concentration and cell density in the pitch. It is also used to look at bottle sediments in the case of quality issues and to generally understand the microbiological makeup of the brewery. Ensure staff members are qualified for primary care of the microscope. In terms of skills, don't assume prior experience is enough. Formal college-level training is preferred, along with experience working in a medical or food manufacturing lab.

Microorganisms in the brewery can be differentiated and distinguished in terms of basic type (shape)

Kohler Illumination

The steps to Kohler illumination are fairly straightforward, but they require some basic knowledge of a microscope. The goal is to achieve uniform brightness and optimal illumination to the sample as portrayed in this graphic. Improperly adjusted microscopes will cause edges of bacteria, yeast, or sediments to be amorphous, and potentially cause inaccurate identification of the substance. For further instruction, please see http://zeiss-campus.magnet.fsu.edu/print/basics/kohler-print.html.

A

B

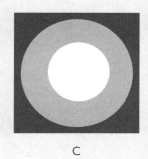
C

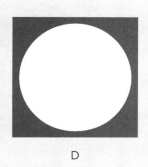
D

1. Remove the oculars.
2. Open the iris diaphragm (under the microscope's platform) 100%.
3. Lower the light source 100%.
4. The field of view will have a bright white light and black ring. Raise the light source until the black ring has disappeared to the edges.
5. Close the aperture of the iris all the way. Do the same procedure as in Step 4 by opening the iris until the black ring has completely disappeared.
6. Replace the oculars and focus on a sample. It should look bright and clear.

There are microscope training schools that teach this basic technique, as well as how to properly conduct other light staining techniques such as dark-field and phase contrast. Even experienced microbiologists can inaccurately identify one type of bacteria for another using standard brightfield illumination. The basic rod or cocci microorganism shape can actually be very difficult to distinguish in some sample types. This could be the difference in reporting a minor infection or a major infection just by inaccurately reporting cell type. The web resources from Zeiss and other microscope manufacturers are also very good for understanding the basics. It is important to take the time to practice this skill with your microscope. Because the light source is often lowered and raised, or the aperture on the iris is opened and closed, it is important to occasionally check this setting. With a properly adjusted light, the edges of microorganisms will be sharp and clear, and not too light against the background. If anything, this technique helps save the eyes of the analyst. Worst-case scenario, counts are misrepresented without proper illumination. Therefore, be sure your training program has at least one person who has been well trained and continues their training with the microscope.

with the use of a microscope. The types of organisms most interesting to a brewery are bacteria (rods or cocci) and yeast. Any brewery of any size can have a microscope since used microscopes can be picked up relatively cheaply. Even the lowest of budgets can squeeze one in, so don't plan to set up a lab without one. The microscope should have at least 40x objective with dual 10x oculars (two eye lenses) for comfort. That will provide a resolution of 400x. For a clearer picture of bacteria, a 100x oil immersion objective is necessary, or 1000x total. The microscope needs to be housed in a fairly low-dust, low-traffic environment. For this reason, it is recommended to purchase a dust cover even if it is in a clean lab; this keeps maintenance and cleaning time to a minimum. The microscope itself can be a very expensive or very inexpensive investment, depending on how many options it has.

A research and development microscope will likely have a camera and several types of light sources, such as polarized or fluorescent. Polarized lenses and other lighting sources are beneficial but not necessary for basic QC work. Buy the least expensive but most reliable model for the budget. Though a microscope can be easy to operate, the microorganisms may be hard to identify if not properly illuminated.

There are several types of illumination the analyst may know how to perform, however, brightfield illumination is the most important and it is the standard used in the brewery lab. With brightfield illumination, the slide is prepared with a wet mount (a wet sample such as yeast) and covered with a cover slip. Incandescent, white light is positioned under the specimen and passes through the slide and sample. With proper focus, the yeast, bacteria, or sediment can be viewed. If the light is too bright or the resolution is too low, the sample view will be poor. A bacteria rod may look like cocci, or calcium oxalate sediment may look like stars, and so on. The technique named "Koehler illumination" helps ensure ideal resolution through the sample (see side bar).

To ensure good microscope technique and microorganism identification, it is recommended to occasionally challenge the analysts to identify positive microorganism controls as part of QA. Positive controls are pure cultures of specific microorganisms that are provided by culture laboratories. Some positive controls are quantitative, meaning they are meant to provide a precise dosage of organisms for plating. Others are qualitative only, meaning they are provided for identification purposes, such as to test Gram staining technique or validate a growth media's ability to differentiate between organisms. Positive control samples can also help test the lab technician's ability to dilute samples, set up the microscope, and perform basic wet mount slides. Many companies sell positive control samples in prepared slides or as dried samples to re-hydrate. Companies that supply these samples can be found on the Internet, though few supply the brewing industry directly. If a microbiological media tests for *Lactobacillus,* for example, any *Lactobacillus*-positive control should grow on it. The sample company should be able to match a control for the test. Other resources to check include the ASBC archives and the ASBC microbiological methods.

If the brewery has progressed its QC program to include Gram staining, an alternative technique that does not require staining is the Non-Staining KOH Test. Also known as the "string test," bacteria can be tested for gram negative or gram positive rapidly. This test requires fewer chemicals to keep on hand in the brewery lab, using 3% sodium hydroxide to determine the gram reaction of anaerobes. This is a simple visual test; a loopful from a single CFU of bacteria, incubated for at least 18–24 hours (or enough time to get a single visible colony) is mixed continuously with 3mm of aqueous sodium hydroxide. If strings or a jelly-like consistency form when the loop is lifted from the mixture, the test is gram negative. Further information on positive controls can be found in the resource section of this book.

Cell Staining
Other tests microbiologists must often perform are cell staining tests. Like the basics of microscopy, it takes a skilled hand to properly perform staining technique, such as the Gram stain. If done

improperly, a spoilage organism can slip by and wreak major havoc. Although it happens to be one of the more finicky tests to perform, the Gram stain is used to identify microorganisms not only by their shape, but by a particular trait in their cell wall make-up—the presence or absence of a peptidoglycan layer (made up of amino acids and sugar). This basic test helps differentiate bacteria types into two distinct groups: gram positive and gram negative organisms.

How does the stain work? Stain gets trapped in the organisms with a heavier cell well. The peptidoglycan layer is very thick in some bacteria types (gram positive) and thin in others (gram negative). This staining technique takes advantage of this difference and will either lock in the first cell stain of blue (in gram positive organisms) or allow the stain to be de-colorized and re-stained red with the second stain (gram negative organisms). Gram positive organisms are classic beer spoilers, while the gram negative organisms are the typical wort spoilers if found at the right levels. Gram negative organisms are also good indicators of a waterborne cleaning and sanitation issue (some vector point is not being cleaned and sanitized well). Therefore, simply differentiating the Gram stain of gram positive or gram negative rods can help in determining a root cause of a microbiological issue (such as gram negative organisms that can cause sanitation issues), but also let you know if you should be concerned about further issues with beer infection. See Figure 6.1 for Gram staining technique.

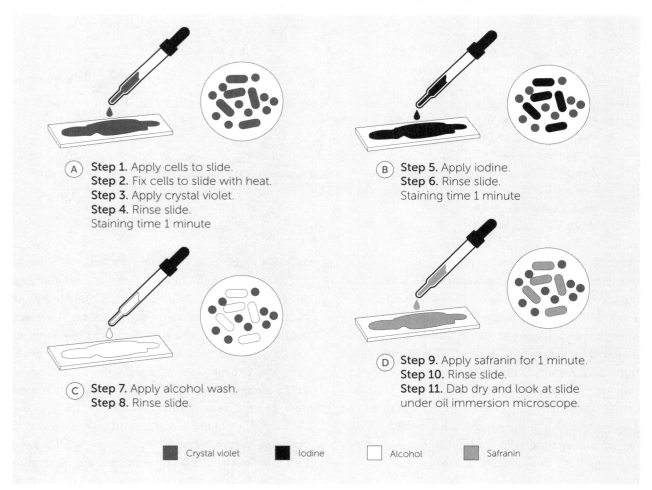

Figure 6.1: Gram Staining Technique – This figure shows the Gram staining technique. In a brewing lab, the Gram stain is used to identify bacteria as potential beer or wort spoilers. Yeast is not stained using the Gram test (it will stain universally purple).

Gram Staining Mistakes

The typical failure modes in the brewery lab during Gram staining:

1. **The sample (of bacteria or yeast from a Petri dish) must be applied thickly, but not too thickly.**
   If too thin, the cells wash off the slide completely. If the sample is too thick, it is hard to see through the layers. The rule-of-thumb thickness level is best determined by placing a slide on a magazine and trying to read through the slide as the slide is prepared. Using a "loop" (a thin rod with a loop end, usually sterile), apply a small sample of cells and allow the slide to air dry. If the analyst can see text, but not read the text while looking through the slide preparation, it is likely thick enough of an application. If the analyst cannot see the black typeface through the sample after it has dried, it is way too thick. This guide is most relevant when a yeast sample also contains some bacteria that are being identified by Gram staining. A thick yeast sample will prevent the bacteria in the sample from being stained.

2. **The sample must be properly affixed to the slide.**
   This is accomplished by quickly passing the slide over an open flame twice. If using an alcohol lamp, pass the slide over the top part of the flame. It will feel slightly warm to the touch on the back of a hand. Overcook the slide, and the cells break apart. For this reason, do not use propane torches to flame slides. The flame source is too hot.

3. **Let the slide cool. Don't rush this part.**
   The slide should be picked up comfortably and no heat should be felt.

4. **Flood the slide with the stains.**
   Don't be too stingy! This goes for rinse water. Rinse with lightly flowing water. A jet of water will wash off the cells, but be certain the stain has stopped flowing from the slide.

5. **The alcohol step can easily be overdone.**
   If this step is overdone, a gram positive identification becomes gram negative. This is the worst false negative a brewery lab can report, as gram positive organisms are the beer spoilers). To eliminate the question, practice on a positive control as a QA check of this test (a confirmed gram positive organism from a certified laboratory).

Other Special Microscopy Tests
Other cell staining tests involve yeast cells. The most critical being the Methylene Blue viability stain test. This ASBC-approved method has been around since 1980. It is a standard test, but is known to have some issues. The test works differently than the Gram stain (Gram stain is not used on yeast, only bacteria). Instead of solely relying on the cell membrane (in this case) to prevent or allow in a stain, the dye passes through the membrane, healthy cells oxidize the stain inside the cell, and thus render it colorless. Unfortunately, budding (or dividing) cells do not perform the oxidation step very well. An actively dividing yeast cell isn't a dead cell, but can be confused as one in this test. Most brewers have been trained to understand this test is only reliable above 90% viability (1997, O'Conner-Cox) and that it is a troublesome test because of this. Other stains, such as methylene violet, and florescent stains that require a florescent microscope, can also be considered as a replacement for the methylene blue test. Each test type has its risks and benefits and requires some consideration to the technique, preparation time, and turnaround time. For this reason, many brewers continue to use the methylene blue stain as a standard procedure. If the yeast culture is healthy and mostly viable (above 95% most of the time), it is a durable enough test to continue. If your culture is more finicky, and may borderline the 90–95% viability level, a stain that is more sensitive should be considered.

Yeast Viability and Vitality

What has been a long debate within the professional brewing chemist and microbiologist community is whether there are assays that not only best predict viability (if the cell is alive) but also vitality (metabolic activity). This long-standing issue has allowed for testing innovations (Trevors, 1983). Some tests require specialized equipment, while others only require the use of a microscope or pH meter. Figure 6.2 summarizes some of the tests.

Note: The ASBC referenced tests are the only tests that are considered industry test standards in the United States. The others may be in use by various breweries as an internal standard.

Viability Test	Reference	Vitality Test	Reference
Slide Culture	ASBC, Microbiology Yeast – 6 Yeast Viability by Slide Culture	CO_2 Generation/O_2 Uptake	Peddie, 1991
Brightfield Methylene Blue or Violet Stain	ASBC, Microbiology Yeast – 3A Dead Yeast Cell Stain (International Method) Smart, 1999	Acidification Power Test	Kara, 1988
Fluorescent Dye (Requires Fluorescent Microscope)	Chilver, 1977 McCaig, 1990 VanZandycke, 2003	Magnesium Release	Mochaba, 1997
		Intracellular pH and Flow Cytometry	Imai, 1994 Bouix, 2001

Figure 6.2: Viability vs. Vitality Tests for Yeast. Vitality tests are considered controversial and are the subject of ongoing research.

In conjunction with viability tests, a technician or brewery worker may also assess the yeast cell density with the hemocytometer. This is a specialized slide designed to count blood cells. It is regularly available in science catalogs for purchase, though it can be expensive to replace. The cover slip is also specialized and must be purchased with the hemocytometer. The accuracy of a cell count with hemocytometer has much to do with the skill and technique of the technician performing the test. The test is performed by taking a thick slurry of yeast from a brink or storage vessel, stirring the sample, and then diluting a small sample down to a thin enough level to count. The first challenge of counting yeast is consistent dilution.

Dilution of any thick yeast sample takes some practice because it can have beer and gas (CO_2) impregnated within it. It requires a consistency in gently stirring the sample in the lab and getting a homogenous slurry. Almost like beating whipping cream, the analyst has to gently mix up the yeast to beat out the gas and ensure a nice, consistent slurry of beer and yeast. Next, the pipetting of the slurry requires attention to detail; sloppy pipetting is a common mistake. A small error in dilution becomes a big error in cell density. In this case, an extra droplet of a thick yeast culture can contain millions of yeast cells. Last, shaking each dilution tube (the 1:10 dilution, 1 gram slurry in 9 grams saline, then the 1:100 dilution, 1 mL or 1:10 into 9 mL saline), requires consistency. A strong repetitive shake, not unlike a bartender shaking a drink, is needed. After proper dilution is made (usually 1:100, depending on yeast thickness), a drop is carefully placed under the cover slip (pulled in by surface tension) and the counting chamber is filled. Be very careful not to go

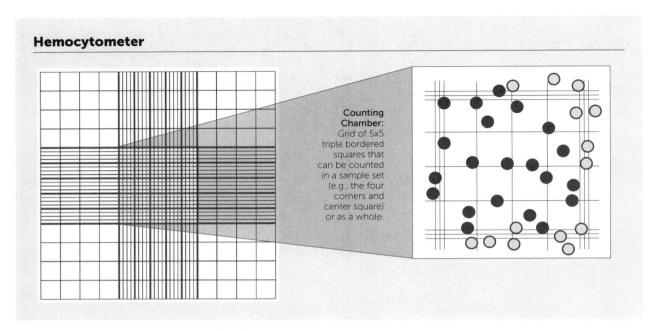

Figure 6.3: Picture of Hemocytometer counting grid.

into the moat, or the well surrounding the counting chamber.

The counting chamber is effectively a three-dimensional area on the slide that has a PRECISE amount of fluid in it. (See Figure 6.3.)

The math of the hemocytometer is not always well understood, but it is fairly simple conversion of a volume in cubic centimeters to mL.

- The counting chamber is
  0.1 cm x 0.1 cm = 0.01 cm^2.
- The depth of the chamber is 0.01 cm.
- Therefore the total volume is:
  0.01 cm^2 * 0.01 cm = $.0001$ cm^3.
- To calculate the number of cells in a mL, multiply the total number of cells in the counting chamber times 10,000.
  Concentration cells/mL = Number of cells total x 10,000.

The tricky math comes when dilution is applied. Most often a 1:100 dilution is used. The concentration of cells calculated from the counting step also needs to have the dilution applied to it. In this case, the concentration should be divided by the dilution applied. This allows the microbiologist to count either the entire counting chamber, or a portion of it (if the cell density is high), and make a calculation to assess the number of cells/mL in the pitch slurry.

This test is one of the harder tests to transition from the laboratory to the operations team. The math and the technique of making dilutions can be cumbersome, but ultimately it is a good practice to have a cellar staff be able to perform this test, and convert to either a pitch weight or volume, especially if this is the only way the brewery is determining yeast cell density in the pitch. Cellar staff should be very familiar with the technique of cell counting because yeast pitching may take place at any time of the day or night, and is best done just prior to the pitch operation.

On the other hand, if cell counting is performed in-line with automated equipment such as radio frequency, turbidity, or other automation, then the validation of that equipment with hemocytometer cell counting is best kept in the hands of the laboratory as a quality assurance test. In this case, multiple replications (at least three different dilutions, counting both sides of the counting chamber) should be performed and the average taken. This QA test is frequently performed at least weekly. However, the challenge is it must be performed for each yeast type used in the brewery as different cell dimensions, and

Hemocytometer Math Example

This is how a cell count in both the upper and lower chamber are averaged to determine the yeast cell count/mL. Note the count between the two chambers has to be within 10% of each other (ASBC Yeast-4 states 10%), otherwise the test is not valid.

1:100 dilution of yeast slurry = 1/100 = 0.01

Total cells in chamber 1 = 750
Concentration of cells/mL = 75 x 10,000/0.01 = 750,000 cells/mL

Total cells in chamber 2 = 800
Concentration of cells/mL = 84 x 10,000/0.01 = 800,000 cells/mL

Average of both chambers = 775,000,000 cells/mL

It is more straightforward to just multiply your average cell count by your dilution factor and then by the area under the coverslip:

Concentration $\frac{cells}{mL}$ = 77.5 x 100 x 10,000 = 775,000,000 cells/mL

Using this calculation will allow weighing out the yeast in pounds based on the target pitch rate. It is one of the single most important tests that any small brewery can do to ensure a consistent fermentation.

---

clumping characteristics can cause slight variation in the automatic cell counting equipment. The frequency of conducting this QA test for in-line cell counting is dependent on many factors. Weekly is most often recommended, but it may be less or more frequent depending on conditions the equipment is kept in and the equipment manufacturer recommendations.

The failure modes that occur during the use of the hemocytometer test include:
1. Wrong cover slip
   a. If the correct cover slip breaks, do not use a standard slide cover slip. The hemocytometer cover slips are properly sized and weighted.
2. Poor filling technique
   a. Too much liquid in the counting chamber causes spillage into the side wells.
3. Too little liquid causes dry pockets under the cover slip.
   a. Poor dilution technique
4. Poor counting technique
   a. <10% variation between chambers (both top and bottom chamber of the hemocytometer)
   b. No more than 15% variation between analysts (two analysts count the same slide, and compare results)
5. Overcrowded counting chamber
   a. A good rule of thumb is no more than 50 cells in any one triple-bordered square, or a total of 250 in the counting chamber.

Microbiological Plating and Media Management
There are many other critical skills and requirements in general that can be missing from a brewing microbiology lab, depending on how much technical training the lab staff brings in. The lab processes of plating samples on microbiological media require specific basic skills that can be easily missed during

training, or conveniently forgotten for shortcuts. The areas we will discuss are tank sampling, pipetting, dilution, media preparation, and autoclave operation, incubation, and membrane filtration.

Sampling from Tanks and Zwickel Maintenance
Tank sampling is probably the first operation that will be questioned (and questioned frequently) if there is a test that comes back positive for microbiological growth. The sample zwickels (as shown in Figure 3.4) in the brewery are notorious for being left in an unclean condition, and if not cared for properly, they can harbor microorganisms. Despite good practice, though, the laboratory technician's technique in sampling is frequently criticized as the root cause of a tank microbiological issue (versus looking into sanitation practices) and possible cross-contamination avenues.

Zwickels can be maintained by ensuring they are part of the clean-in-place (CIP) procedure, and routinely pulled from the tanks and re-built (disassembled, fully cleaned, and re-greased), and are not overheated during the sampling. Some zwickel designs are very different. Be sure to review the design specifications with the manufacturer, as some can be steam sterilized as opposed to alcohol sanitized. The typical process is carefully spraying the inside and outside of the zwickel opening with 70% lab-grade ethanol, and then applying a quick flame from a traveling propane tank to ignite the alcohol. This practice can be improperly done, and thus cause some issues with the zwickel cleanliness. Be sure the lab analyst or staff has NOT been trained to heat the zwickel until it is red hot, which will leave a scorched and charred zwickel. This practice will not maintain the zwickel in good condition. Instead, it may cause false negatives or false positives due to caramelizing of beer on the zwickel causing a harborage point. Or, the sample may be drawn from too hot of a sample port, killing all organisms in its wake. Instead, remember the zwickel was designed to be cleaned gently with sanitizer, flushed, sample taken, and sanitized again.

Proper zwickel valve sampling technique:
1. Spray 70% alcohol directly into the valve outlet. Spray the outside of the valve. If preferred, use a sterile swab dipped in 70% alcohol to scrub and clean out the valve outlet first.
2. Ignite the alcohol. If you desire, a propane flame torch can be used. Simply light the alcohol quickly and allow the alcohol to *burn off*. The only reason to use a flame is to fully remove the alcohol. However if an open flame is a safety concern, plan to keep the valve extra clean, spray with sanitizer, and flush it properly (Step 3) to take a sample.
3. Flush the valve for a 10–20 second period at high flow (it is ideal to completely flush out the sampling line before collecting a sample). This step is critical and many times not done.
4. After the flush, slow the flow down and take the sample by quickly passing a sterile tube or other container under the flowing beer. Do not touch the outside of the zwickel with the test tube.
5. Aseptically cap the tube and close the valve.
6. Finish the process by thoroughly spraying the zwickel down again with alcohol, inside and out. A swab can be used again here to clean out the valve. Some zwickels can be capped off with sanitizer left in them as well.

An alternative to maintaining multiple zwickels on every tank is to "valve off" (using hygienic design) each sample port with a tri-clamp valve, and then port a zwickel to the tank with clamp and gasket. If this is done, it is typically asked for during tank design. This is very safe as the zwickel and gasket can be soaked in sanitizer between tanks. The other advantage of this method is that the sample site (a valve opening) can be very easily cleaned and sanitized with a brush just before and after sample taking. The valve can also be left packed with sanitizer and capped off. The downside of this method is that it can be time consuming. The primary consideration is to keep the sample method simple, maintain a hygienic zwickel condition, and use common sense. If common sense is utilized, the zwickel should not be your source of microbial tank issues.

Eliminating the Zwickel Sanitation as Root Cause

The zwickel and sample technique can indeed be a source of bad sanitation, and the root cause of a positive count on a plate, but this is frequently not the root cause. To eliminate the zwickel as the source, the first question to ask is, "What microflora are on the plate?" If it is exceptionally varied, including wild yeast, molds, and bacteria, then there may have been an issue in tank sampling from poor zwickel cleanliness, or from not using sterilized containers to transport the sample. If it is a single colony, or only a few types of organism, then it is *highly unlikely* the root cause was sampling. Lastly, a dirty zwickel or poor sampling will not only show an exceptional variety of microorganisms, but the organism types are usually aerobic growers, not anaerobic. If the plate has single organisms, beer or wort spoilers, or a single type of wild yeast, then do not challenge the sampling technique. Start looking into sanitation processes as a potential root cause.

Pipetting and Other Challenges with Microbiological Plating

Once the sample has safely made it to the laboratory, the analyst still has plenty of work to do, mainly keeping the sample clean, cold, and sterile until it is plated. One rule of thumb is that the sample should not remain un-plated any longer than 24 hours, even if it is kept in the refrigerator. An active fermenting sample will change in that period of time. Therefore, some brewing labs may have stricter rules. If a sample is held for too long, and it is fermenting, the viable cells from the gram negative bacteria families may die off, thus giving a false negative. These are the organisms that can show poor sanitation and should be an indicator to review sanitation procedures, rinse water, or other investigation. False negatives at this stage can lead to spoiled beer.

Typically the analyst will plate the sample on several different types of growth media, and may even incubate the media in aerobic or anaerobic conditions. This helps cast a wide net on the types of microflora that may be present in the sample. The plates may be pour plates, spread plates, or membrane filtered, depending on the sample type. The media selection is up to the brewery to determine, and their known biological issues. A summary of media can be found in Figure 6.4. Consult the media jar for proper incubation temperatures for each type. Most breweries can get away with one 30–32°C incubator and can use an anaerobic jar (a small jar that has a pouch inserted for anaerobic incubation). *"Microbiology Methods in Brewing Analysis"* is a good reference for further guidance (Campbell, 2003). It is recommended to review this media list yearly to be certain that the assays are still relevant, or if process improvement has allowed risks to be lowered.

Next, the lab analyst has to dose the sample, cleanly and precisely, onto the media plate. Using sterile technique to pipette the sample also takes practice, as the analyst is applying a vacuum to the pipette and drawing out the carbonation, which causes havoc when attempting to measure a precise dose. The colder the sample is, the better. Sterile antifoam in the tip of the pipette can help with exceptionally stubborn foaming beers; however, that may impede precise volume delivery of the sample, so use judiciously. Precision of volumes is important, though not as critical as chemistry tests. In microbiology, the difference of 10 cells on a plate versus 20 cells could be natural variation or error. Therefore, good technique is recommended, but don't lose sleep over exacting precision. If you want to compare things over time and apply a dilution factor, then precision is important. A droplet may contain quite a few microorganisms. If the analyst is partially emptying the contents of a pipette, then precise delivery of the volume is expected, without touching the contents in the tip.

For a pour plate, once the samples are pipetted onto the sterile plates, the tempered media (45°C, no hotter) should be poured onto the plate off to the side of the sample and swirled immediately. Several pour plates can be lined up and poured consecutively, but if a sample waits longer than 30 minutes for media, a false negative situation may occur. Best practice is to keep the time from pipetting to media pouring fairly

Media	Type of Plate	Incubation Type	Type of Sample	Organisms
Plate Count Agar	Pour	Aerobic	Water, Environmental Swab	General
Universal Beer Agar, Wallerstein Differential or Nutrient (WLD, WLN)	Pour/Spread/MF	Aerobic or Anaerobic	Beer, Water, Wort	General, Beer Spoilers, Wort Spoilers, Yeast Spoilers
Lee's Multi-Differential Agar	Pour/Spread	Aerobic	Beer	Beer Spoilers, Wort Spoilers
Raka-Ray Lactic Acid Bacteria Medium	Pour/Spread/MF	Aerobic or Anaerobic	Beer, Yeast	Beer Spoilers
Lin's Wild Yeast Differential Medium	Spread	Aerobic	Yeast	Wild Yeast
Lysine Medium	Spread	Aerobic	Yeast	Wild Yeast (Non-*Saccharomyces*)
Hsu's *Lactobacillus* and *Pediococcus* Medium	Pour	Aerobic	Beer, Wort, Yeast	Beer Spoilers
Selective Medium for Megasphaera and Pectinatus (SMMP)	Pour	Aerobic	Beer, Yeast	Beer, Wort, Yeast Spoilers
DeMan Rogosa Sharpe Medium	Pour/Spread/MF	Aerobic	Beer	Beer Spoilers
Barney-Miller Brewer Medium	Pour	Aerobic or Anaerobic	Beer	Beer Spoilers
X-α-GAL Medium	Spread	Aerobic	Beer	Ale vs. Lager Yeast

Figure 6.4: Microbiological Media in Brewing Laboratories (Partial List) Source: ASBC Methods of Analysis, online. Table of Contents, *Microbiology Methods*. American Society of Brewing Chemists, St. Paul, MN.

short, so plan accordingly. If pipetting onto a spread plate, spread the sample immediately, as it will sink into the agar and grow any colonies in one area only.

The quality assurance tests that microbiologists perform include hood maintenance as well as testing for negative and positive controls. The sterile hood is essentially a large air filter that passes sterile air over the working space. A sterile hood is not essential to perform simple microbial testing, but very helpful to prevent mold growth. Any plate with spotty mold growth may indicate air contamination, or the hood filters may need to be replaced. Replace the HEPA filter as indicated in hours of use. When new media is prepared and autoclaved, there are several QA checks to perform. The microbiologist should test the media for a negative control (pour a plate and do not add sample) for every batch autoclaved. This requires the media be plated after removal from the autoclave to ensure it was sterile. Every batch of media should be tested for pH and results documented on a running log. Once the media is autoclaved, and QA tests performed, it can be cooled and refrigerated for later use. Date and number these lots appropriately. A microwave can be used to melt down media, however, if melting media down, be careful not to scorch and overflow the media from the jar. For this reason, only fill the media jar 2/3 full to leave room for bubbling while heating and cooling.

When ready to use a media for plating, melt it down via microwave or direct fire (a lab Bunsen burner, for example, being careful not to scorch) and cool it in a water bath to 45°C. It can be held in the water bath at 45°C for an extended period of time, but longer than eight hours is not recommended. It is beneficial

Hsu's *Lactobacillus/Pediococcus* (HLP) Test

Very small breweries need not be intimidated by microbiological test procedures or their costs—this test is a great place to get started and will open the door to other tests that can add value to your small brewery's testing program. The Hsu's Lactobacillus/Pediococcus (HLP) test is very simple and can show you that not necessarily all microbiology tests need to be done by a trained laboratory technician. The important thing in quality is to find a starting point and this is as good as any—there is a wide variety of these kinds of tests that small breweries can use in the quest for quality.

Designed at the Siebel Institute (the JE Siebel and Sons) by Dr. Hsu, HLP media detects *Lactobacillus* and *Pediococcus,* the two most commonly found lactic acid bacteria in the brewery. It is a selective media (meaning it selects for specific bacteria types, and prevents others) and has the added benefit of low-cost preparation. It requires boiling the media in a flask and then pipetting in sterile test tubes. It can be stored in a refrigerator and then warmed in a water bath prior to inoculation. 0.1 mL to 1.0 mL can be inoculated in the tubes. HLP contains Actidione, a chemical that suppresses yeast growth, and an oxygen scavenger in the media. It is a semi-solid and considered a semi-anaerobic media.

For a visual representation of HLP media, the Siebel Institute has a video available on YouTube and the specific method is found in *ASBC Microbiological Control-5.*

General overview for using HLP media:
1. **Prepare the media.**
   a. Weigh out dry HLP media. Follow the package instructions for water to media ratio (7 grams media to 100 mL water, unless making a stringer solution). Pour media into a sterile flask that can contain at least 2x the liquid amount you will be making and that can be heated in the microwave or on a hot plate. Add the water and mix the media well by shaking the flask gently.
   b. Heat the flask gently, in either a microwave or using a hot plate, until the media has boiled at least two minutes with the cap loosely in place.
   c. Cap tightly and cool in 45^0C water bath.
   d. With sterile pipette, transfer 15 mL media into sterile 25 mL test tubes. 15 mL is approximately all that is needed. Cap using sterile technique (don't touch the inside of the tube or cap with your hands). The media can be stored in the refrigerator for a week, or used immediately. If the media is cooled below 25^0C, be sure to warm it with a water bath prior to next step.
2. **Inoculate the media.**
   a. Once the media is at room temperature, pipette your sample (0.1–1.0 mL) into the media. A good technique is to stab the media with the pipette directly to the bottom of the tube and slowly lift and dispense the inoculum. Another technique is to pipette the sample into the tube, cap it, and gently rock the tube (no bubbles).
3. **Incubate the media and record results.**
   a. Let the capped, inoculated test tubes incubate at $28-30^0$C for 48–72 hours. Check for any growth.
      i. If growth occurs, gently scrape a colony from the media, place onto a slide, and use wet-mount or air dry, heat fix, and Gram stain to confirm type.
      ii. There are some differentiation qualities to the media. Lactobacillus can appear as narrow comets, while Pediococcus appears puffy. Still, it is recommended to verify any growth under a microscope.

> **Training and Laboratory Manuals**
>
> Though this is an overview of media and preparations, nothing beats actually doing the tests and having a trained microbiologist initially certify skills. It is highly recommended to get training prior to starting a laboratory. There are also many short training courses available to learn the techniques described here. In addition to training, having excellent print references in the laboratory is helpful. The lab manual most consulted in the US brewing industry is *The ASBC Methods of Analysis*, now available online. This reference is the standard for the US industry. However, there are other resources on the web to help get a very small brewery started with microbiological testing, including a free manual from the Brewing Science Institute, *Brewer's Laboratory Handbook*.[1]

as part of QA to use a positive control organism to challenge any media with a new manufacturer's lot code. As mentioned previously, positive controls can be purchased from various companies that sell pure culture stock of microorganisms. These are usually in freeze-dried pellets and can be very easy to rehydrate. Select the organism that you want to be sure will grow on the media. (For example, if the media is selective for *Lactobacillus*, use a *Lactobacillus* as a positive control.) Rehydrate as directed and plate the sample. Incubate as normal for the media. If the positive control grows as expected, the media is good to use and the lot can be recorded as passing this test. If the positive control does not grow, check the media pH again, and contact the media manufacturer for help. This lot may be suspect and should be set aside until a positive control test is passed.

As mentioned in the sidebar on hemocytometers, dilution of yeast cultures is particularly difficult and requires special discussion. Yeast from a propagator or brink is thick, sometimes paint-like, and sometimes very buoyant and "fluffy." It is critical to properly stir the yeast in larger samples, or shake vigorously if contained in a sterile jar. Do not use a stir plate—it will heat the sample and possibly kill the yeast or microorganisms. Some labs require a certain count of seconds of stirring or shaking to standardize this process. However the process is done, it is best to standardize it and ensure all lab analysts use the same technique by verifying training. Assuming the yeast is homogenized, the analyst will carefully pipette and dilute it. If not working aseptically (not plating the sample on media), a wipe can be used to wipe off excess yeast on the outside of the pipette so excess is not transferred. Some labs opt for weighing the yeast on the first dilution for consistent transfer of 1 mL or 1 gram. After the first 1 mL is diluted into 9 mL of saline, the sample should be vigorously shaken. The sample is ready for the second dilution of 1 mL into 9 mL to make a 1:100 dilution. Take caution if pipetting 0.1 mL into 9.9 mL to make the same 1:100 dilution. Although this seems to be a more efficient process, it may result in more variable test results for yeast samples due to trub and other particles that overtake the volume. Also, it is very difficult to consistently transfer 0.1 mL of a thick slurry. If both 1:10 dilutions are properly treated and shaken, and this operation is done consistently between analysts, the test results will be highly repeatable and reliable. However, if an analyst is rushing and skips the vigorous shaking, tries to cut corners, or gets a little sloppy in pipetting, results can be more variable. For that reason, take the time to properly train all staff that perform a yeast slurry dilution.

CHEMISTRY TESTS

The brewing lab typically has a large dedicated area that is chemistry focused. In general, most of the chemistry tests that monitor the basic critical quality measures can be performed with little investment and still provide the brewery a high level of accuracy and precision. Of all the tests performed, alcohol is one of the key chemistry tests. Other tests for color and bitterness use "wet chemistry" (mixing beer with various chemicals and color assays) and a spectrophotometer. For performing titrations for tests such as calcium

---

[1] http://www.brewingscience.com/PDF/BSI_brewers_lab_handbook.pdf

or titratable acidity, a burette is needed. Measuring pH requires utilizing a pH meter. However, brewing labs can also utilize more advanced tools such as the high performance liquid chromatography (HPLC) for measurements related to hop compounds, gas chromatography (GC) used mostly for yeast and fermentation related compounds, and atomic adsorption spectrophotometry (AA) or Inductively Coupled Plasma (ICP) for metals analysis. Many of these instruments require specialized training and skills along with specific requirements for electricity, lab temperature, ventilation, and additional gases. They also can be exceptionally pricey and out of the reach of smaller breweries. If an instrument such as a gas chromatograph is purchased, it is best to engage the original equipment manufacturer to provide detailed maintenance guidelines and advice on how to best manage beer samples. Therefore, this portion of the chapter will focus on the basic wet chemistry tests, and the nuances of beer measurement.

Background on the Chemical Methods Related to Beer

As the governing body for beer chemistry, the American Society of Brewing Chemists (ASBC) methods have been the go-to guideline for most brewers since the group's inception. ASBC is still a great networking body of chemists and microbiologists. This chapter references the ASBC Methods of Analysis, 14th Edition online, though there are other organizations that provide guidance to brewers for testing, including the European Brewers Convention (Analytica-EBC) that can be referenced.

Measuring Extract and Alcohol Background
The term "extract" in brewing relates to all fermentable and non-fermentable sugars (carbohydrates), nitrogen compounds, salts and minerals, and other substances that make up what is extracted from the mash/lauter operation into the kettle. Although the wort extract is not 100% carbohydrates (90–92%), the scale used to measure density of extract in brewing was derived from pure sucrose solutions. We use the density measure to relate back to a percentage of sugar solids in the wort, usually in the unit of Plato. Density is mass per unit volume of a substance (kg/m^3). To simplify calculations, a dimensionless or unit-less ratio such as specific gravity (SG) is typically used. The SG is the ratio of two densities: the density of the substance (wort or beer) over the density of pure water at 20°C. The determination of SG or extract density is used to estimate the concentration of extract in the liquid in degrees Plato (°P) by utilizing conversion tables. The original tables devised for brewing to relate SG to % extract in g/100 g were developed by Balling, and were for pure sucrose for 100 g of solution SG at 17.5°C. Later, Plato corrected some inaccuracies in the Balling tables. For this reason, percent or °P is the more accurate measure of extract percentage in worts and beer. Although these tables were originally constructed at 17.5°C, for purposes of brewing the variance from 17.5°C to 20°C in SG measurement did not introduce enough error to reconstruct the tables (1958, DeClerck, 15).

The original gravity is defined as "the specific gravity of the wort from which the beer was fermented" (DeClerck, 1958, 426). Measuring extract becomes more convoluted after yeast is introduced, however, because the alcohol interferes with the actual reading (alcohol being lighter than water). Fermenting beer extract is called "apparent extract" because, as previously mentioned, the alcohol interferes with an actual sugar content reading. "Real extract" (RE), sometimes called "residual extract," is the actual extract in the sample. The relationship of a beer's original extract, the resulting alcohol, and residual extract was originally tested and experimentally determined by Balling in 1943 (DeClerk, 1958, 426). Over several experiments, Balling was able to predict how many grams of extract in °P would give rise to 1 g of alcohol. Through a series of precise experiments, Balling would evaporate the alcohol and CO_2 from a 100 g sample of beer, and re-weigh the residuals extract. Balling determined 2.0665 g of extract resulted in 1 g of alcohol, 0.9565 g of CO_2, and 0.11 g of yeast. Since CO_2 and yeast are removed from the beer, each gram of alcohol is equivalent to 2.0665 g extract. Also, since 100 g wort does not elicit 100 g beer, the equation also takes into account the weight of CO_2 (0.9565 g) and yeast (0.11 g) in the denominator.

$$p = \frac{(2.0665A + n) \times 100}{100 + 1.0665A}$$

p = Value of original extract in grams per 100 g or Plato
A = Alcohol by weight
n = Grams of extract

With this knowledge, Balling created a method to calculate the original gravity of a sample based on the final gravity, despite the alcohol interference. He also made it possible to calculate the alcohol if the original extract and residual extract are known. The tables became the basis of all alcohol calculations in brewing. Most brewing labs are not measuring RE in fermenting samples. Instead, they are taking a direct density reading of the fermenting beer and utilizing the AE component in the formula to calculate alcohol and OG content.

Measuring Extract on the Brewdeck
On the brewing deck, a hydrometer is typically employed to measure wort extracts or validate in-line instrumentation of extract. The hydrometer is an accurate instrument that is quick and easy to use, therefore ideal for a small brewery. However, the accuracy and repeatability can vary more than 0.2°P if not reviewed for calibration. Hydrometers work in a fundamental principle of buoyancy of a "float" and will vary depending on the density and temperature of the liquid. The hydrometer floats up for higher density liquids, and sinks in lower density liquids. The scale should read to a certain level of accuracy, and must contain a thermometer to make the temperature adjustment. Purchasing a weighted hydrometer in the beer scale of Plato with a mercury temperature scale has become increasingly difficult, though. Some states no longer allow mercury thermometers and the alcohol thermometers are less reliable. Brix, as mentioned previously, is a close representation of Plato and can be used instead. Some hydrometers read in SG, which is fine, but SG is cumbersome unless all recipes are written to formulate with it. The best advice is to use the hydrometer in the scale the brewery uses that delivers the most accuracy and precision. Regarding quality assurance, calibrate the hydrometer with a sucrose solution against a more precise instrument in the lab every six months. If the hydrometer measurement is off, "correct/calibrate" the hydrometer by making a notation to add or subtract a certain amount of degrees Plato. This should be done with all thermometers in the lab.

Some key factors in reading a hydrometer accurately:
1. The hydrometer must be DRY.
2. The liquid needs to be as close to 20°C as possible.
3. The cylinder must be level.
4. Gently lower the hydrometer in the liquid to the approximate level and allow it to float; do not drop it. Allow 2–3 minutes of wait time before a reading is taken.
5. Read just below the curve of the meniscus at EYE LEVEL.
6. Check accuracy of hydrometers frequently with a standard laboratory sucrose solution, especially at the points of the scale most likely used.

Refractometers are also employed on the brewdeck for a quick read of extract. These work by refracting a specific light wavelength through the sample. This test is also a temperature-sensitive measurement. Refractometers can be purchased for many types of liquids, and sugar solutions measured in Brix are common. As refractometers are sensitive to particles, bubbles, and temperature, and are only available in the Brix scale, they are not recommended. However, if a portable refractometer is used, and it provides good accuracy and precision, then by all means continue to use it.

The original method of determining extract in the laboratory required precise use and weighing of a dry small vial called a pycnometer. The first step is to weigh the empty, dry pycnometer, then weigh it with exactly 25 mL or 50 mL of 17°C distilled water. The pycnometer is then emptied, cleaned, and dried, and exactly the same volume of 17°C extract is added before it is weighed again. The ratio of the weight of extract to water is the specific gravity (SG).

The use of pycnometers to measure SG is laborious and slow, but it is a precise method of determining specific gravity. Thank goodness it is no longer needed!

There have been many advancements in measuring extract density in the lab with instrumentation. Today's brewers use more precise instruments in the laboratory, primarily the U-shaped oscillating density tube or "cell," originally designed by Anton Paar in 1967. This instrument measures the frequency of the oscillations that pass through a U-shaped tube. Along with temperature readings, the instrument calculates density. The U-shaped cell is less accurate when bubbles are present. Also, temperature and alcohol must be compensated for. Handheld density meters with a U-shaped cell are handy for smaller labs, and can be used on the brewing deck instead of the hydrometer (Anton Paar, 2015).

Alcohol measurement has also evolved over the last 50 years in brewing. Methods have ranged from distillation to ceramic sensors that pick up volatiles from alcohol oxidation (SCABA), to the current technology using near infrared (NIR). NIR spectrometers work by collecting and analyzing light in a specific infrared region on the light scale that is passed through the sample. The detector picks up on the spectral absorption of light related to its molecular vibration and compares it to the light that is not absorbed by the vibrations (2015, Anton Paar). The percentage alcohol is determined by this difference. This technology is used to measure blood alcohol levels and percentage alcohol in spirits, wine, and cider, in addition to measuring beer. Despite being a bit pricey, it is a very robust and accurate method that many breweries find worth the investment because it is fast and very easy to use. If just starting up, a brewery can still estimate alcohol by using the extract calculations. Since current instruments can be fitted with a near infrared (NIR) reading of alcohol as well as a U-shaped density tube, a very precise extract value is achievable.

If direct alcohol measurement is not available to the lab, distillation, or simply calculating alcohol content from the Balling equations, is still a good method. However, the Balling equation does not compensate for error due to volatile acids in sour or *lambic* style beers. Volatile acids have a higher specific gravity than water. If sour or *lambic* style beers are being made, more precise alcohol measurement by near infrared reading, or by modifying a distillation method with caustic, is advised (1958, DeClerk).

Spectral Analysis

Another instrument used frequently in the laboratory is the spectrophotometer. This is used for many direct and indirect colorimetric assays such as color, international bitterness units (IBUs), polyphenols, free amino nitrogen (FAN), sulfur dioxide (SO_2), and others. The spectrophotometer has become a fairly inexpensive tool for most labs to analyze for IBUs. A spectrophotometer capable of analyzing at a lower wavelength can increase the cost of the instrument, though it can add more test capabilities. As with most lab instruments, a temperature-controlled, dust-free environment is necessary to prevent equipment failure. The instruments are typically very easy to use, allowing some pre-programming of calculations, making the read-out in ppm or IBUs, etc. The instrument operates on a simple principle that light at a certain wavelength is passed through a sample of a specific length and the resulting light that passes through or is absorbed is used to calculate a value. The tests usually run on a spectrophotometer are color and IBUs. (The spectrophotometer operates under the principles found in the Beer-Lambert Law of physics.)

Some cautions in spectrophotometry:

1. Be sure the sample is clear of any sediment and does not have bubbles. The light MUST pass through the sample to be detected at the other side, and bubbles, sediment, or yeast will all interfere with measurement.

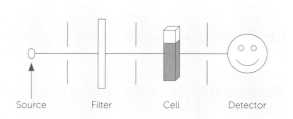

2. The cell can fog if the temperature is very cold and the lab is very warm. Overly colored (dark) samples must be diluted or use a shorter path length cell.
3. The cell must be very clean and dry. Traces of previous samples, water, or other residue in the cell interferes with measurement. One commonly used method is to rinse the cell with the next sample and then fill it for testing. This

flushes the residual of the prior sample out, and ensures an accurate representation of the current sample.
4. Most tests have high and low limits. Be certain to know if your sample is close to the limits, and then reduce any error that can be introduced.
5. All tests have accuracy limits. Do not assume the lab is at fault if the test gives variance in data. The test and analyst may be performing as well as expected. To find out what level of accuracy is tolerated, consult the ASBC original method, or participate in the ASBC check sample service.

Measuring pH

pH measurement is a critical measurement in brewhouse performance and one of the easiest tests to bring into a brewery. It is commonly used for everything from checking water chemistry and mash pH to ensuring final wort pHs are to target to fermentation (a rapid drop in pH is required for a healthy fermentation). It is even important in cleaning chemistry (e.g., making sure residual chemicals have been rinsed properly). Depending on the technology being used, there may be sources of error in this test, but they can be easily controlled with some awareness. pH is a measure of the active hydrogen protons in solution. It is calculated as the negative logarithm of the hydrogen ion and measured in mol/L. For example, a pH of 7.0, which is neutral pH, is effectively 1×10^{-7} mol/L of $[H^+]$ (hydrogen ions).

The types of pH meters and probes are vast, but the technology is fairly simple in concept. Two electrodes (one reference, one glass) measure the total potential change in millivolts when a sample is exposed to a reference ionized gel (salt bridge). The electrical potential that is measured across the glass membrane of the pH probe tip is a result of the difference of electric charge between the two liquids. This charge is equated to pH. This is all completed in fairly rapid time and is a repeatable method of measurement. All pH meters have to be calibrated routinely. Calibration with two points (pH 4 and 7, for example) allows for a slope of a line to be calculated. Monitoring this slope will help determine if there is a decay in the probe, either in the reference solution or the electrode itself (Kohlmann, 2003).

Other obvious issues with pH have to do with staff awareness of their method. A sample that is too hot or too cold will not measure accurately, no matter how good temperature compensation is. A good practice is to bring the sample as close to 20°C as possible. Failing to rinse the probe, scrubbing it with a wipe (instead of gentle dabbing), or simply being too gruff with it will cause the sensitive electrical measurements to become off kilter. Lastly, allowing the probe to dry out, get crusty with solution, and not keeping it in fresh buffer will cause malfunction. If this is the case on the brew deck, pH papers are a convenient means to rapidly assess pH. These are not ideal for brewing liquors, but can be used in a pinch or if pH probes are consistently not maintained in the brewery. pH papers are best used for checking the pH of rinse waters and cleaning solutions.

SENSORY ANALYSIS

Sensory analysis, or the scientific discipline of tasting, smelling, and describing beer, is a critical QC tool to utilize in the brewery. Ideally, all beer is tasted by a trained staff prior to it being released, during filtration, at the final blended tank, and in the package during packaging. One of the hardest decisions in a brewery is to stop production and not package a tank that is ready for packaging because it "doesn't taste right" or has an off-flavor. This usually upsets plans in packaging with personnel and sanitation scheduling, and further downstream it can upset deliveries, scheduled truck routes, and many other issues. For some, it seems incredible there are any failures at all at the final tank blend that hadn't been detected by other microbiological and chemistry checks. Yet there are many opportunities for failure that possibly only a sensory check can detect. For example, the finished beer may have been cross contaminated with other beer of a similar alcohol and color, incorrect levels of sanitizer used and higher residuals remained, or incorrect spices used in the kettle for a spiced beer. In other words, some failure modes at this step are ONLY detected by a trained sensory panel, so it is a critical check to perform.

To conduct sensory analysis as a quick QC check in any phase of the brewing process, there are some key questions to ask.

1. Is the staff properly trained to conduct the check?
2. Are the conditions proper to conduct the sensory test?
3. What standards are go/no-go?

We will explore each of these below.

Taste Panel Selection

The ASBC has an excellent protocol to follow related to the selection and training of a taste panel. It requires screening processes, and testing each taster to see if they can detect off-flavors. Once the panel is selected, the tasters are routinely exposed to off-flavored beer and continuously trained. Though these methods are excellent, not all breweries have the luxury of having a pool of personnel to choose from, let alone the hours it requires to train them. Still, tasting has to be done and in a small brewery it is possible the brewmaster may not be available for every situation in which a go/no-go decision must be made. If this is the case, selecting members for a taste panel comes down to who is available. The person or persons doing the work to prepare a tank or filter the beer can perform a taste panel screen in every situation. They need proper coaching and training, though. In other words, if the brewery is short on staff and cannot perform the steps to bring a properly trained taste panel up to speed, start small. The recommendation is to use the human resources you have to conduct sensory analysis and train them on the specific failure modes of the process.

To train on a budget, it is recommend to walk through all the failure modes of the process with the trainee and make it real using a sensory check. For example, if you are training on proper sanitation techniques, the training would include showing the staff how to test sanitizer levels. It is also recommended to ask the lab to prepare a sample of beer that was artificially adulterated with the flavors that excess sanitizer would bring on (such as oxidation or chlorophenols), and have staff members smell the sample. They will quickly see how much a small error can impact the beer. If the lab does not have these flavors, then adulterate the beer directly with a sanitizer or cleaning chemical, and only smell the beer with the trainee. Have them learn to use their senses on what true-to-brand looks like, feels like, and smells like by purposely showing them "the bad" with off-flavors. This can be done with other areas of off-flavors, such as over dilution. Rather than preparing a tank totally full with the improper dilution water amount and only finding out with analytical tests the tank was batched incorrectly, have staff members instead try a sample with proper versus improper dilution. Depending on the process, if dilution is in-line, the staff can always taste a sample on the way to the tank to ensure the ratio of blend water is correct (though this taste test may require too sensitive of a palate for many). If the possibility of dilution water is left in packed lines, filters, etc., have the staff test the first beer coming off the filter at the tank. If adding flavors, try a beer with incorrect flavors added and so on. This type of training may not be all encompassing to every off-flavor; still, training on directed off-flavors at the point of failure is just as effective and important.

Simple off-flavor tests to train without using purchased off-flavors:

- Caustic or sanitizer in beer – DO NOT DRINK THESE. In 100 mL of beer, add 1 mL of caustic or sanitizer. Smell the beer and note the off-aroma.
- Diluted beer – Dilute a brand by 10%. Taste it and note the differences in bitterness, mouthfeel, etc.
- Diacetyl – A common issue in small breweries. To train for this flavor, some equipment is required. Heat a water bath to 60°C. Take a small sample of actively fermenting beer in a Nalgene bottle or glass jar with cap that has no aroma (fill the jar half full). Hold at 60°C for one hour. This will convert all the potential precursor flavors to diacetyl. Remove cap and smell. This may require a finished "control" beer to smell alongside the sample. The beer will smell "hot," old, oxidized, etc., but if diacetyl is present, it will smell like movie theater popcorn.
- Yeast issues – Yeast from the brewery can be brought into the lab and "heat abused" by holding in a hot area for 12 hours. Smelling the yeast fresh vs. a decaying sample is a great way to train for a sensory check of pitching yeast.

After this type of training is performed, the staff should feel empowered to stop a process if the beer or yeast tastes off. If the process is stopped, other trained staff will need to verify if the beer is correct before continuing the process. This verification sensory test needs to take place in a neutral area, away from production with other staff members that were trained in the off-flavors, as well as a manager with professional off-flavor training. Taste the questionable sample against a standard, doing a blind test if possible.

Additional Sensory Tests

For verification purposes, the best test to perform is either a duo-trio or triangular test. These are tests that add some level of objectivity to verifying whether or not a sample is to standard. They require a trained staff member who understands how to choose which test is more appropriate, how to conduct the test, how to analyze the data statistically, and how to interpret the results. They also require at least six people to participate. Alternatively, simply tasting two samples against one another objectively works. If setting up a duo-trio or triangular test, refer to Figure 6.6, and refer to the ASBC Methods of Analysis for more detail.

Discrimination Tests

- **Question:** Are two products different from one another?

- **Triangle Test:** Choose the sample that is most different from the other two samples.

- **Duo-Trio Test:** Choose the sample that matches the reference

Both tests answer "Are two products different from one another?"

Figure 6.5: Triangular and Duo Trio Tests. This figure shows the triangular test setup and the duo trio test setup for taste panel.

As suggested above, not every brewery will be set up in a separately designed panel facility for a full taste panel to perform routine go/no-go or other sensory checks. Many sensory checks are performed at the source of the quality check. For example, a filtered beer may be tasted immediately as filtration begins to ensure no dilution of off-flavors, or the yeast may be tasted and smelled prior to pitching. It is important to still follow some points to ensure any taste or sensory check is being done in a place that allows for full focus on the task. As with any quality check, if sensory is being performed during a process, be certain all tools, cups, and sampling vessels are kept in a clean environment, away from outside contamination, and that proper sampling sites (clean, sanitary valves) are in order. Once the sample is taken, be certain there is a neutral odor area nearby that can be used to smell or taste the sample. As convenient as it is to conduct the check on the line, or at the filter, etc., the sensory part of the check may be impeded if the staff isn't focused on the task due to other smells in the area that are troublesome. If a separate small room is available, use that. If the ideal conditions are not present, be sure to test the responsible staff member's ability to pick up the off-flavors in that environment. Provide adulterated beer or samples to check this acuity. Lastly, make sure each test has a process to follow if the test fails the first sensory check. If the keg line steam does not pass sensory, what's next? The production staff may have a manager conduct the test and ask for assistance from the boiler maintenance technician to be called in if the test fails a second time.

If using a sensory panel to conduct go/no-go tests, then be certain the panel has a centrally located, quiet, and odor-free place to meet. The panel room can be a set of booths in a hallway, or a round table in a small conference room. Panel rules and the standard go/no-go criteria for every beer should be posted. Talking should be minimized during the test, as should any noises, cell phones, scoffs, and laughing. This is a fun activity, but it needs to be objective. Scoring sheets need to be available and easy to understand. Though it is tempting to change the template for a scoring sheet, resist this to avoid overloading the panel. They need to focus on taste, aroma, visual appeal, and mouthfeel. For cross-training purposes, it is helpful to discuss the beers as a group when the panel is completed.

Sensory Analysis Test Kits

Test kits are available from a wide variety of sources (see Resources). The brewery will pay for the convenience of these kits, but it is well worth the cost because purchasing the raw chemicals from proper chemical supply houses instead can be costly for delivery (many are flammable and delivery costs reflect that). Additionally, storing and diluting the chemicals can require training. If the samples are not properly spiked, it can injure the tasters.

Some key points to consider when ordering a kit:
- Test kits come in a couple of different forms, liquid and capsule.
  - The liquid forms are highly flavor active and dissolve easily in beer, but have to be micropipetted to be delivered properly to the beer. These can decay quickly once opened.
  - The capsule forms can be harder to dissolve, but easier to measure. These also have a slightly more stable matrix. Some brewers complain that the capsule format aromas are not an exact match to the actual off-flavor in beer; therefore, the risk is a panel that can be trained on flavors that won't actually exist in beer. However, these kits improve every year.
- Have a plan to use up the kit. It is expensive, so know what flavors need to be utilized in different applications, and use the kits for extra training as needed.
- Store the kit as recommended. Do not use beyond the stated shelf life.

Criteria for Go/No-Go

The go/no-go full taste panel or quality-at-the-source sensory check will need clear criteria and guidelines for panel members and staff to use. These guidelines have to be clear, concise, and direct. A good set of go or no-go criteria may also be brand specific and involve not only off-flavors, but a range of flavors that are considered critical for the beer, such as hop aromas or yeast characters. These can be called "on-flavors." The panelist then assesses the level of on-flavors and is also asked to be certain there are no off-flavors in the beer. An example ballot is in Figure 6.7.

The trick is setting the no-go for the levels of on-flavors and the more obvious flaws of off-flavors. For example, if one panelist thinks hop character is too strong for the brand, but no other panelist's ballot agrees, then it may pass. If two reject the hop flavor, then the beer may be set on hold, a triangular panel called, and a final decision determined (blending or running as is). A quality at the source check should be focused on off-flavors only. The reason being that work-in-progress beer, or individual components such as hops or malt, are difficult to assess if they do not meet a specific level when smelling or tasting as a quick check. Instead, use more objective means to determine if a raw material does not have a character trait. As an example, if a hop from a specific region does not have a characteristic piney aroma that was expected. The hops may still need to be utilized, as they are not off-flavored, but if a true-to-type hop character is missing, it is best to ask the supplier to check on the sample. They may run more sophisticated checks to determine if the hops were cross blended or labeled incorrectly.

Once taste panels or sensory checks of any kind have been initiated, the brewery must coordinate keeping the people on the panel in check and calibrated, just as any other sensitive instrument in the lab. This taster validation requires, at the very minimum, exposure to off-flavors on a frequent basis using test kits. A rigorous example is a monthly blind tasting run with six samples of a selection of off-flavors where the panelist is asked to identify them. The panelist must maintain a score of 66% or better to remain on the panel. This rigorous taster validation may be required for larger breweries that require highly sensitive palates. An example of a less rigorous validation is conducting a blind off-flavor test once a quarter. Another idea for continued calibration of the panel is to validate them against one another by using a competitor's products

Time:	Name:

Name of Beer

Color and Appearance
☐ Appropriate ☐ Not Appropriate ☐ Dark ☐ Light
Comments: _____
_____

Aroma
☐ Appropriate ☐ Not Appropriate
Comments: _____
_____

Bitterness
☐ Appropriate ☐ Not Appropriate ☐ High ☐ Low
Comments: _____
_____

Alcohol
☐ Appropriate ☐ Not Appropriate ☐ High ☐ Low
Comments: _____
_____

Flavor and Aftertaste
☐ Appropriate ☐ Not Appropriate
Comments: _____
_____
_____
_____
_____

Mouthfeel/Body: Very Light   Light   Medium   Full

Balance and Drinkability
☐ Appropriate ☐ Not Appropriate
Comments: _____

Carbonation
☐ Appropriate ☐ High ☐ Low
Comments: _____

Technical Quality
☐ Excellent ☐ Very Good ☐ Good
☐ Acceptable ☐ Needs Improvement
Comments: _____
_____
_____
_____

Figure 6.6: Taste Panel Ballot—An example of a sensory score sheet from Alaskan Brewing Company. It was developed based on the Great American Beer Festival® score sheet.

as samples, and scoring them in the regular manner. This helps make sure the panel agrees on thresholds and levels of flavors.

If a full taste panel is being utilized for a go/no-go check at the finished tank, a key panel validation is to occasionally check the panel's ability to "fail" a beer. Add an adulterated sample that should fail a routine panel QA check every month or two. It keeps the panel alert to the task and also educates them on the process to fail a beer. This seems counterintuitive, as one never wants beer at this point in the process to actually fail. Still, it is less of a risk to train the panel to ensure they can perform the task of questioning a beer's quality at this stage than for the panel to pass a beer that should have failed. If a beer never fails, and the panel is presented with an actual off-flavored beer one morning, the likelihood of failing that sample is low just because it is out of the routine. Therefore, keeping the individuals trained on off-flavors is important, but also keep the panel tuned as a team that has practiced for the worst-case scenario.

PACKAGING TESTS

The challenge of packaging beer is that it is the last stop in the brewery before the consumer sees the product and it presents new failure modes. Packaging poorly can change the beer flavor by improperly gassing or adding oxygen. Or more simple errors can occur such as poor label application. This may deter a customer from buying the product. Damage to flavor from a poor crown application or can seam seal can occur during packaging. It is well known in beer quality circles that packaging poorly can ruin weeks of hard work in the brewery.

Many traditional checks of finished beer may only require visual assessment. These include visual assessment of the package. Typically, breweries pull 1–2 cases per shift or every 4–8 hours to look at several variables:

- Is the label on properly?
- Is the code correct?
- Is the crown on straight?
- Is the diameter of the crown correct (use a gauge to measure the crown diameter)?
- Is the glue of the mother carton applied properly (check the bead and the application)?
- Are the fill heights within legal tolerance?

> **Check the Fill Level, Either by Weight or by Height**
>
> Fill level by weight requires a routine calibration of the package tare weight. Glass, in particular, can change weight over time (even if it is from the same manufacturer). It is best to take an average of 20 bottles roughly every month to validate the tare weight as a QA check.

Carbon Dioxide Checks

The other items checked include the gassing of the beer. These require testing equipment. If carbonation is too low there is a perception of flatness; if it is too high there can be filling issues as well as perception of over foaming. Typical measurements are beer headspace, air, dissolved oxygen in the package, and total package oxygen (TPO), which is a combination of headspace and liquid dissolved oxygen. Additionally, gassing levels and measurement of CO_2 are taken routinely. Determination of beer gas, both CO_2 and dissolved oxygen, is typically performed with a meter or handheld device at the bright tank, and again in the package during the time of filling. The failure modes for oxygen ingress and improper carbonation increase as the beer is finished and processed, so it makes sense to measure at both the finished beer tank and in the package. The greatest opportunity for failure is during the start-up of a packaging line or during beer changes when large gaps between brands may result in dissolved oxygen spikes. Typical sampling procedures logically focus attention at these critical times. At the bright tank, a gassing check can be made just prior to a tank being released to the packaging hall. At the filler, either at the very start of the filler or immediately following a beer change, the brewery collects a random sample of several packages and assesses for oxygen (headspace and dissolved) and CO_2 levels. In both scenarios, any immediate reaction to high or low levels of CO_2 or high dissolved oxygen can be performed (such as stopping a fill, and checking all pipe connections to ensure a tight seal). For this reason, these checks are best performed at the source by personnel that can immediately act on the data.

Saving a set of samples for a QC analyst to check after a couple of hours will result in erroneous readings. For example, if a high oxygen level is measured in the bright tank or in the package, waiting only allows the levels to go down. The reaction to excessively high dissolved oxygen at the bright tank may mean a wasted tank, or that the tank is turned into kegged beer only and put on a fast track for quick rotation. Since most breweries don't allow levels to exceed 200 ppb, the dissolved oxygen levels at the bright tank can also be plotted and watched carefully for trending. If the data is trending, the staff can react by checking and tightening all connections, or opportunities for oxygen ingress at filtration. Along the same vein, CO_2 levels that are too low or too high at the bright tank can be immediately addressed before packaging if caught in time.

Traditionally, CO_2 testing devices for beer are based on Henry's law, which states that at a defined temperature, the concentration of a gas is proportional to its partial pressure at equilibrium. This assumes the pressure and temperature of the liquid the gas is dissolved in is known. Also, it is assumed the vessel has reached equilibrium. In a manual unit, this is achieved by shaking the container filled with a liquid sample vigorously. (There are automated units that operate on Henry's law that do not require vigorous shaking.) If operating the manual unit, follow these steps:

1. Cool the entire testing device. If the chamber that receives the beer is warm, dissolved CO_2 will break free.
2. Fill the chamber off a zwickel and remove the Zahm-Nagel unit.
3. Record the liquid temperature.
4. Shake the chamber until the pressure gauge reaches equilibrium.
5. Convert pressure and temperature to volumes of CO_2 using the CO_2 chart (Figure 6.8).

An inherent challenge with this type of measure is the volume of trapped air or oxygen that is included in the total pressure. Therefore, it has to be assumed the oxygen level is nominal, or if using an automated instrument, there may be an adjustment for oxygen.

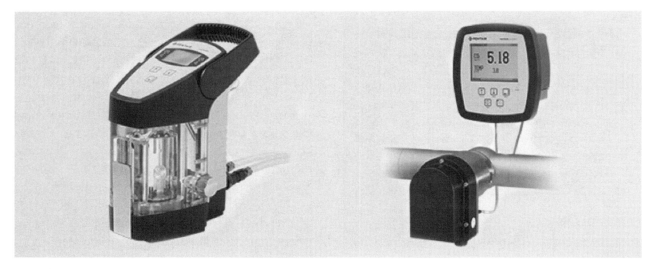

Figure 6.7: Manual Offline Vs. In-line CO_2 Testing Equipment for Measuring CO_2 at the Bright Tank. Image courtesy of Pentair Haffmans, Netherlands.

Also convenient to understand, the measurement of CO_2 is typically reported in percent volumes (Vol/Vol) in the United States, and requires knowing the temperature and pressure the liquid is in. European convention may measure CO_2 in g/L, which is roughly two times the percent volume. For this conversion use the data related to carbon dioxide at Standard Temperature and Pressure. The density is 0.122 pounds per cubic foot, or 1.94 grams per liter. Therefore, multiply % volume by 1.94 to equate to g/L (Kunze, 1999, 579).

Guidelines to prevent basic errors that can occur in CO_2 checks:

1. *Ensure both the pressure gauge and the thermometer are properly calibrated.* This measurement relies heavily on temperature, pressure, and consistency in treatment of the sample. Automated technology relies heavily on the equipment cleanliness and calibration. Manual or automated, if the equipment is off in temperature by even one degree, the CO_2 volumes will be off slightly. Therefore, maintain the pressure gauge and thermometer on these instruments, and keep them calibrated. If using a manual meter, though, it is required to cool the instrument before use, though storing the instrument cold will eventually damage the thermometer. Keep the thermometer calibrated and checked.

2. *Check that the manual device is cooled to the temperature of the liquid.* The entire device needs to be kept as cool as the beer that will fill it.

3. *Bring the sample up to pressure equilibrium in the device.* In a manual measurement, the sample must be shaken VIGOROUSLY. This is awkward and requires some strength. When the gauge no longer moves, it's a good rule of thumb to keep shaking it twice as long as it took to get to that point.

4. *Don't allow the sample to warm while taking the reading.* Don't take so long that the sample actually begins to alter temperature. Use best discretion to get the reading fairly quickly (Zahm and Nagel, 2014).

Other considerations when measuring CO_2 using the automated units that operate on the same principles of Henry's law are that they may vary slightly from unit to unit because the algorithms to process a CO_2 concentration are not standardized (Hutchinson, 2012). Also, any microbiological leaks in the device will cause an erroneous reading. Newer technology is attempting to address the issues that the Henry's law–based devices inherently have. For example, thermal conductivity (TC) measures gas diffusion and thermal conductivity change over a gas diffusion membrane and attenuated total reflectance (ATR) is based on infrared spectroscopy. Understanding

THE BEST TESTS FOR A BREWERY

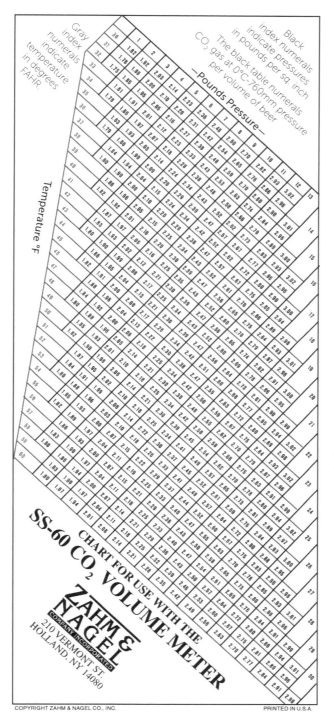
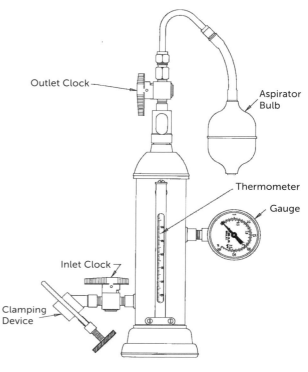

FOR OPERATING INSTRUCTIONS CONSULT THE QUALITY CONTROL TESTING INSTRUMENTS MANUAL.

HIGHER TEMPERATURE AND PRESSURE READINGS ARE SHOWN ON THE CHART IN THE QUALITY CONTROL MANUAL.

Figure 6.8: CO_2 Solubility in Beer and Pressure/Temperature Chart. Permission from Zahm-Nagel. This table is used to determine solubility of CO_2 in beer when using a manual CO_2 measuring device.

CO_2

Type of Instrument	Risks	Benefits
Henry's Law (Pressure/Temperature Devices)	Requires the instrument be kept cold prior to reading, and gauges calibrated.	Simple to operate. Can be lower cost.
Thermal Conductivity (TC)	Measures gas diffusion and thermal conductivity change over a gas diffusion membrane. Requires membrane upkeep. May have issues with delivery over the membrane.	Can use a standard solution to calibrate.
Attenuated Total Reflectance (ATR)	Based on infrared spectroscopy. Doesn't always correlate with standard temperature/pressure readings due to mathematical model differences.	Measures CO_2 adsorption selectively and is less influenced by outside variables.

the technology and algorithms that the technology is using is critically important when performing a sensor check with an alternative technology as part of a QA program. If a brewery has brought in a newer sensor, work with the original manufacturer to understand calibration, maintenance, and quality assurance methods to check readings.

Measuring Dissolved Oxygen
Dissolved Oxygen (DO) is typically reported in parts per million (ppm) in wort and parts per billion (ppb) in beer. Depending on the instrumentation, these levels will be reported in ppm or ppb. As very low oxygen levels can be achieved, it is best to purchase an instrument that provides the highest level of precision for the price afforded. These instruments are normally purchased separately. There are a couple of ways to measure oxygen in a package; via a simple chemical test, membrane, or other technology. Headspace oxygen can be measured using a chemical test with the Zahm Nagel. The chemical test for headspace oxygen requires an additional burette attachment to the Zahm Nagel device. The burette is effectively upside down and has a weak caustic solution in it. After the device pierces a bottle, can, or sample taken from a tank, a valve is opened and the gas pressure is bubbled through the caustic solution. The CO_2 dissolves, while any air or oxygen residual displaces the caustic. The challenge with this traditional method is the headspace air levels may be so low that it doesn't register an actual value. This method measures headspace oxygen only. To measure dissolved oxygen, a separate device needs to be purchased.

The devices that are used to measure dissolved oxygen in the liquid portion of the packaged beer are more costly than the Zahm instrument. They traditionally use membrane technology that allows a liquid sample to flow over the membrane. The oxygen reacts with a sensor and makes a signal that results in an output. These units require a connection or rig that allows a sample to flow over a membrane either from a bottle, can, or the bright tank. The membrane technology also requires calibration at atmospheric pressure, and consistent upkeep. Membranes get worn out, become fouled by yeast, and need a technical hand to replace. Keep these issues in mind when planning and purchasing a membrane device. New technology is based on amperometric or electrochemical technology, and optical measurement technology. These may be in-line or offline. Because in-line meters are exposed to cleaning fluids, be sure the technology is clean-in-place tolerant and that it is low maintenance. Offline testers usually have a piercing device and some way to "push" (usually with nitrogen gas) the beer over a sensor.

Failure modes exist even with the newer technology. First, the package must be shaken very well if the intention is to test for total package oxygen (TPO). If the brewery is interested in determining where a high TPO reading may be coming from, a quick check is to test a sample unshaken and get a quick read on the liquid dissolved oxygen. The difference between shaken and unshaken samples provides a fairly good idea if the oxygen is mostly headspace related or liquid. The next issue regarding reading consistency usually has to do with sensor maintenance, cleaning,

DISSOLVED OXYGEN

Type of Instrument	Risks	Benefits
Burette/Caustic	Requires chemical. Can be cumbersome to shake the package. Doesn't measure very low levels.	Easy to purchase and use.
Membrane	Requires upkeep of membrane. Delivery to membrane must be consistent flow. Can be slow to read very cold samples.	Not too hard to use, and can be cost effective.
Amperometric	Requires zero check and calibration.	Calibrates in air. Fast response time.
Optical	Can be higher cost technology.	Low drift, fast response.

or delivery consistency (the gas pushing the beer has a leak somewhere). These instruments are not inexpensive and require some expertise in-house. They are not recommended to bring into a brewery QC program until it is clear who is primarily responsible for general daily upkeep, and monthly maintenance. They also need to be checked against a standard. This becomes tricky in the brewing industry. The best standard is carbonated water made to a specification of dissolved oxygen. Larger breweries are able to make in-house DO standards. Smaller breweries have to rely on reading a larger brewery's beer, which is typically very well controlled in terms of dissolved oxygen. This is risky, as reading another brewer's beer isn't a standard, but it can provide a sense of whether or not the instrument is way off in measurement. Below is a summary of instrumentation for in-package gas of CO_2 and dissolved oxygen.

Though this chapter discusses a wide variety of tests, procedures, and quality assurance validation issues in QC measurement of beer, the processes can all be standardized and become a very solid heartbeat for the brewery to operate with. Do not get too discouraged if a test is showing higher variation than is desired, or a test result was a false positive. Just as in the brewery, the lab should be working diligently to reduce variation and standardize processes. It is an evolutionary journey.

KEY TAKEAWAYS

1000–15,000 BBL Brewery	15,000–150,000 BBL Brewery	150,000+ BBL Brewery
• Basic lab tests still require good QA. • The lab will grow with the brewery. Make space.	• A formal lab is being built. Bench space and areas to expand microbiology tests should be planned. • Lab skills are becoming more sophisticated with the acquisition of new equipment.	• Microbiology, chemistry, and packaging are all well-developed labs. • Lab skills still being developed. • R&D may take some foothold in the lab area.

SEVEN

GOVERNMENT AFFAIRS

Regulatory pressure on the brewery can be perceived as either extremely onerous or fairly straightforward, depending on who you ask. The regulatory environment includes compliance with federal food safety rules, excise tax laws, and even local wastewater or other environmental rules. As a brewery expands its distribution area, complying with federal rules and new state alcohol label laws becomes more complex. Most of these rules are very similar to the requirements found in any food or beverage company. This chapter will focus on the food safety laws and the measurements required for excise tax reporting. It is written from a quality lens, not a regulatory lens; that is to say, be certain to validate any procedures with appropriate council if in question.

FOOD SAFETY AND RISK ASSESSMENT

Food safety rules have slowly become more formal and all encompassing in this country since September 11, 2001. That act of terror led to broad sweeping concerns over food safety and food defense policy and regulation. The Food and Drug Administration (FDA) began to look again at all food and beverage producers, and realized there were some gaps in the system that they needed to bolster to ensure even small producers were not at risk, or putting the public health at risk, due to poorly designed control systems. This resulted in several layers of food safety control that started with requiring all food plants to register in a national database, improving traceability rules and food defense recommendations, and incorporating a risk-based system in all plants through the new

Food Safety Modernization Act (still under review). These changes were in addition to complying with one of the oldest rules on the books, the Current Good Manufacturing Practices (GMPs), which is required of all breweries, though the issue is debated.

Whether a brewery is a food plant is something that came into question when the FDA first brought about the requirement to register plants with the passing of The Bioterrorism Act of 2002. The FDA was always expected to enforce compliance for food safety, even in breweries; however, most breweries were most familiar with the Tax and Trade Bureau (TTB) and not the FDA.

Even prior to 2002, if a Class I recall were required for a beer product because it posed significant hazards to the public health, the FDA would be the arm of government that was alerted and consulted. Alcohol producers have what are considered low-risk facilities (their products are at a low level of risk for becoming adulterated and posing potential harm to the public health through processing); therefore, most breweries have few interactions with the FDA, or even state government agencies that act as a regulatory body for the FDA. Each state and municipality is different, but even a local health department may inspect the brewery for health code requirements upon opening and return on a routine inspection schedule.

Food Plant Registration
The Bioterrorism Act of 2002 expanded the FDA's authority and further carried the concept of "food defense," thus requiring, for the first time, that every food and beverage producer register with the FDA and show their ability to trace any raw material from receipt all the way through to the delivery of finished goods. This included breweries, and for some, the requirement of tracing every lot of raw materials, such as hops, through to finished goods was a new burden to overcome. Other breweries already had audits and inspections by third parties due to requirements by larger customers or because they made organic beer. These breweries were in a better position to comply with the Bioterrorism Act, as they were already accustomed to many of the same quality and food safety systems, such as food defense and lot traceability. All breweries had to register on the FDA secure website as a food plant, which required approximately 10 minutes of inputting data.

History aside, the requirement for food plant registration has not gone away. In fact, the FDA now requires food plants to keep their registration current, updating records every two years. This is still a fairly simple exercise to do. The more difficult component of the law remains keeping diligent records of raw material lot codes and having the ability to trace every raw material to its finished good lot. Many small breweries can do this on paper only. It is recommended to test this system every year with a mock traceability audit. (See Appendix A for a traceability record example.)

GMPs – A Foundational Requirement
Good manufacturing practices (GMPs) are a set of guidelines imposed by the FDA as part of 21 CFR part 100 that require sanitary production of all food products. Some form of a GMP rule has been part of the United States's landscape of food and beverage manufacturing since 1906. This set of rules covers employee hygiene, building and facility sanitary operation and maintenance, pest control, and the defect action levels in food (such as bug parts or mold levels). The rule is part of the code of federal regulation (21 CFR) that explains the United States Food and Drug Administration's authority and rules. The US FDA may inspect a brewery for its compliance to this law, but they usually rely on local or municipal health inspectors to conduct a routine GMP inspection. If a local code exists, it should be used to influence your GMP policy (see Appendix H) and the hygiene design of your facility. This GMP policy is part of your quality manual. Because of their relationship with large customers, your supplier of hops or malt is another resource for this type of policy. These policies should be reviewed and adopted in each brewery, specifically for the risks they carry. For example, specify if hairnets or another type of head covering (such as a hat) are to be worn, and exactly where in the facility they are required. Simply adding a brewery name to an existing document, but then not complying with the policy, is more harmful than having no policy in place.

How to Conduct a Traceability Audit

1. Select a raw material, such as a hop, and note the lot code. Pull all production records into which this hop lot went. This is the target raw material.

2. Add up the total weight received, and the weight used of the target raw material from the production records. There should be 90–100% retrieval of the target raw material on paper. If more is missing, or cannot be found, it is important to determine how or why and make corrections to the process for future productions, either in receiving or use records.

3. Next, make a note of all the finished beer lots the target raw material went into. This becomes the target finished product. There may be many lots of production related to one lot of raw materials. This is common. What also is common is blending productions together into one tank, making a new lot code. For both of these reasons, this part of a traceability audit can be time consuming. Diligent record keeping, or a well-designed inventory and production system, can make this a much easier task.

4. Pull all the shipping records of the target finished product. Again, the brewery should be able to trace 100% of the target finished goods to the distributors involved. A good invoice or billing system will allow for lot codes to be included as part of the invoice. This can make looking for multiple codes an easier task.

5. Keep a record of what the target raw material was, percent recovery of the raw material and the target finished goods, and any process corrections made because of the audit. This record may come in handy for a larger customer seeking evidence that a system is in place, or for a government inspector looking for a brewery's diligence of their record keeping and traceability system.

Once GMP policy is in place, the next step is implementation of the policy. The requirements seem relatively simple to understand and follow, such as removal of jewelry or not wearing pens in front pockets. However, because these rules require a change of habits, GMP policy can be notoriously difficult to establish and maintain.

The golden rules of any change implementation for GMP policy are:

1. Plan to communicate the policy in simple terms first and allow time for the change to take hold. Train staff on the *what* and *how* of the policy, as well as *why* the change has to happen.
2. Allow employees to be part of the change by bringing up concerns or questions. If a part of the policy is not set in stone, talk it through and change things to make the policy easy to follow.
3. Give employees time to learn the rules, and allow for a transition period. If there are a lot of new rules, establish them slowly—say, one rule a week.
4. Communicate progress and reward the employees as they anchor the change.
5. Keep GMPs part of new employee training.
6. Establish the same protocol with office staff, sales staff, and visitors. Any dilution of the policy will result in less adherence by employees.

Once policy has been implemented, the brewery leadership should routinely verify policy adherence. Either a casual walkthrough while taking note of the facility, employees, and maintenance practices that need to be buttoned up or more formal audits and documentation can accomplish this. Self-inspection breeds a culture of policy adherence.

If all of the above is in place there shouldn't be cause for alarm if the brewery is approached by a local health department inspector or a United States FDA inspector for an annual GMP inspection, as the brewery is following its GMP policy. Ask to see their badge, provide them a copy of the visitor GMP rules, and then the inspector has a right to inspect the brewery. Assume the inspector will see the brewery as a lower-risk facility, and will not plan to take up days for their visit. As mentioned previously, breweries are known to pose little risk to the public health due to sanitation concerns. However, occasionally an inspector will be new, or decide for whatever reason they want to see much more than a quick walkthrough. The inspector will know there are enough other general areas to review to be sure the public health is protected, even in low-risk breweries, and they will likely want to find something. Keep this in mind, and have an open conversation with the attitude of learning, and they will usually be very kind in the end.

The areas of high interest for an inspector are as follows:

1. **Employee hygiene**

    A brewery must provide employees with tools at their disposal to keep their hands and equipment clean. Depending on the municipality, this will likely be with a hand-wash, a 3-compartment sink, or both. Signage, hot water, and paper towel disposal will be of concern to the inspector. Hand-wash sinks, either in the bathroom or in the general facility, show there is a process and policy for hand washing. Be sure that signage is visible and the policy says when hand washing is necessary.

    The inspector will want to make sure the employees know when to come to work, and when not to, in case of illness. The policy related to employee health and employee hygiene should be well known by all employees. If an employee is sick, are they sent home? Are employees allowed to wear shorts, smoke, or eat anywhere in the facility? If an inspector sees evidence that an employee isn't following what is stated policy, they will note it on their report.

2. **Facility and equipment sanitation**

    The inspector usually looks for signs that the facility or equipment is not being cleaned and sanitized properly. On brewing equipment, any rust or poor welds would be noticed, as would heavy mold or pink yeast. In the packaging hall, be certain no overheads are dripping condensate over clean bottles. The cleaners and sanitizers used on equipment should be approved for food contact (Menz, 1974). Be sure the documentation that states this (it may already be on the cleaner's label) is easily accessible. Secondary containers to carry cleaning and sanitizing chemicals, such as buckets, should be labeled. Also, equipment sanitation records should be easily accessible. Be sure to document sanitizer residuals on these records if a non-rinse sanitizer is used.

    In the facility, any infestations of bugs, open doors to the outside, rodent droppings, or other clear physical deterioration is likely to be noticed and written up. Also, any hop buckets, malt bags, or tools stored on the floor or on an unclean surface will raise red flags. Keep these on a pallet or cleanable surface.

3. **Maintenance practices**

    The last broad category an inspector will review is general maintenance practices. Specifically, they will look at boiler chemicals (are they food grade?) and lubricant control (are food-grade lubricants labeled and does a policy exist that is clear on where to use them?), and to see if any part of the facility is in gross disrepair. Some areas that may draw their attention are the spent grain removal areas (these can be smelly and poorly maintained), grain receiving hookups (these should be locked, secured, and clean), and condition of the mill, silos, and warehouse.

A good policy, well-trained employees, and a diligent effort to keep the brewery tidy and organized will impress any visitor, especially an inspector. Keeping the GMP policy current, training frequently, and making it part of the everyday routine of brewery management is the best practice.

Fill Level and Alcohol Monitoring
The Tax and Trade Bureau (TTB) requires adequate fill records for a brewery, as well as reporting the total production volumes and alcohol levels for tax purposes. For this reason, following good practices on calibration, record keeping of equipment, and validating procedures keeps the regulatory pressures to a minimum

Food Safety Modernization Act

On January 4, 2011, President Barack Obama signed the Food Safety Modernization Act into law. The full implementation of this law is still pending at the time of publication, however, the key areas the FDA wanted to improve upon were preventive controls for animal feed and food safety for humans. The animal feed proposed rule may have overly burdened breweries by establishing a HACCP-like protocol for spent grain. The FDA was still considering this portion of the rule at publication. See the current standing at the FDA's website or the Brewers Association website (BrewersAssociation.org).

if in an audit situation. The TTB's website is an excellent reference to what is commonly found in audits and how to avoid issues: http://www.ttb.gov/beer/beertutorial.shtml#_Production_and_Inventory_3.

Monitoring tank fill levels with a sight glass or an in-line meter requires some sort of calibration and validation. Keeping a record of this validation is necessary for the TTB to be satisfied. Larger breweries may routinely swap out in-line calibration meters with certified services. Bottle or can fill levels are a measure the TTB will be interested in. The best methods are gravimetric, however, beer weight and container weight can vary. If using the gravimetric method to calculate container volume, be sure the container tare weight is updated every quarter. For example, don't assume the weight of your glass bottle 10 years ago is the same today. Take an average weight of 10–20 bottles every 2–3 months as the tare weight. If measuring volumetrically (the liquid must be room temperature for accuracy), the displacement of the liquid by the gas has to be accounted for or eliminated. Beer should be de-gassed and warmed in the lab prior to taking the measurement. The ASBC Methods of Analysis (http://methods.asbcnet.org/summaries/fills-1.aspx) has specific procedures for taking fill-volume measurement.

Alcohol measurement by direct measure needs to be calibrated with a standard such as a reagent grade ethyl alcohol. These reagents can be purchased and diluted down to a level near beer (5%, 7%, 10% by volume) and measured on the device the brewery uses. Any calibrations or adjustments should be documented. Additionally, the brewery can and should routinely send their samples out for third-party validation of test data at an independent laboratory. This is especially true for the packaging breweries that service large restaurant chain accounts due to the 2014 ruling on nutritional labeling requirements in restaurant chains. Alcohol, nutritional content, and calories will need to be validated by a third party on an annual basis.

The government requirements on breweries are many and involved, but most have to do with interstate commerce and tax laws. Regarding food safety, the brewery needs to maintain a clean and well-organized brewery and be able to track and trace production, as in all food facilities. Keeping these policies and procedures simple and to the point will help make certain they are followed.

KEY TAKEAWAYS

1000–15,000 BBL Brewery	15,000–150,000 BBL Brewery	150,000+ BBL Brewery
• GMPs are stated in the quality manual and followed. • Logging lot codes should be a regular occurrence. • A standard procedure to pull product out of the market is on record.	• GMPs become formal, plant-wide. • Brewery has lot code traceability records. • Brewery has planned recall procedures, and has performed traceability audits.	• GMPs are formal; all employees, visitors, and contractors are well apprised of rules. • Brewery has an ability to trace all raw materials and finished goods. • Brewery has planned recall procedures and performed traceability audits.

EIGHT

PULLING IT ALL TOGETHER – ASSESSMENT TIME

Every two or three years the brewery and quality leaders should step back and assess the brewery's overall quality systems to determine if any processes are lacking and whether the brewery is appropriately spending its resources on the areas of most risk or concern. This is called a quality audit or assessment. The concept of a quality audit can feel a little overwhelming to operations staff who are busy making, packaging, and shipping beer. (Indeed, if timing or the process seems off-putting, then it should be carefully considered if the assessment itself is a good fit for the brewery!) Still, a good quality system audit has a clear objective to verify that the quality policy and procedures are being followed and that the policy and procedures are the proper "fit" for the brewery. Over time, the product quality risks that a brewery adds will grow in scope or severity. Concurrently, the resources to perform the quality procedures will sometimes shrink or get cut. The audit results will tell the brewery management team if the quality system is being followed and if it is equal to the risks the brewery has assumed for the product mix at that point in time.

> The audit results will tell the brewery management team if the quality system is being followed and if it is equal to the risks the brewery has assumed for the product mix at that point in time.

In this chapter we will introduce the concept of the quality system audit, why it is important to perform, and what some audits entail. The goal of this chapter is to help the brewery ensure the audit structure it selects is not an over-reach of reviewing minute details, but instead results in a clear assessment of what needs to happen in the next one to three years in order for it to keep moving its quality system forward.

TYPES OF AUDITS

Before we delve into the why of conducting an audit, we first have to differentiate between the types of audits that are typically performed.

This chapter will focus on the Quality System Audit (system audit), however, conformance and compliance audits may come into play during a brewery's lifespan and are worth a mention here. Conformance audits are directly related to a certified quality system such as ISO or Safe Quality Foods (SQF), where standards, criteria, and requirements of documentation, management structure, and the quality system are well defined. These audits are less common in smaller breweries, but may take place as part of a corporate oversight to large brewery conglomerates. A third party that is certified by the agency (such as ISO or SQF) usually performs this type of audit.

The compliance audit is related to criteria and standards that are required by a state or federal regulation for good manufacturing practices, labeling, and/or the weights and measures. A third party from the state, such as a health department, usually performs this audit. The TTB will also perform audits for weights and measures related to fill heights and month-end reporting documentation. A compliance audit ensures the brewery is compliant in all legal requirements related to public health and record keeping for tax purposes.

The quality system audit is the most common and comprehensive audit performed in a brewery. The brewery management team, a quality manager, or a third-party consultant may perform this type of audit. Large or small, this can be a fairly heavy undertaking when one considers all of the components, including:

1. Process Assessments – This determines if quality and/or operations procedures exist and are being followed according to the SOP. The process assessment is the bulk of a quality system audit.
2. Department Assessment – This reviews both training effectiveness and who is doing what related to the quality system.
3. Product Assessment – A review of the product itself and its performance in the field (data review or consumer complaints). This data can be used to fix any QC/QA procedures that may allow defective product into the market. The data can also be used to assess whether a product can conform to a specification or whether a specification may need to be re-considered. Examples of these three audit components can be found at the end of this chapter in figures 8.1, 8.2 and 8.3.

Even a mid-size brewery's quality documentation can contain hundreds of pages of policy, procedure, and specifications. Add in a product assessment review and it may take over a month to conduct the audit properly. Some larger breweries may split the audit into categories such as laboratory, packaging, brewing, and utilities for ease of review and resources. While this process entails much time and effort, there are many good reasons to invest in an assessment of this magnitude.

The first reason to conduct an audit is that audits can prevent a slow decay of product quality by taking a broad view of process, product, and risks. Some breweries wait until a quality system breaks down and causes a major failure, either in the field or at the brewery, before realizing this purpose and initiating audits.

Another reason to conduct audits is best practices and money-saving opportunities can be discovered; a good audit should ultimately find places where money can be saved.

For example, one brewery found wild yeast in a cellar whose cleaning procedures became inadequate after several key employees retired. Through trial and error, the retired employees had learned that only a specific water line brought good flow from the CIP station because an engineering change further upstream caused certain pipes to reduce flow. After the personnel and process changes occurred, the beer from that cellar was blended with other batches. Over time, the occasional consumer complaint of "bandaid" flavor started to pop up. In this case, it wasn't until a quarterly assessment for wild yeast that the problem was caught. An audit may have allowed the brewery to find the issue with water pressure and change procedures before the failure occurred by revealing the cleaning and sanitation SOPs were different in this cellar than others. Instead, this brewery determined only after a failure occurred that the quarterly microbiology test for wild yeast did not adequately address checking for wild yeast during a change in engineering. An audit process may have prevented this failure and possible loss of customers.

Another reason to conduct audits is best practices and money-saving opportunities can be discovered; a good audit should ultimately find places where money can be saved. Tests or layers of process are sometimes added to a brewery's processes due to a short-term correction of a problem. For example, a brewery had a temporary issue with filter breakthrough leaving sediment in the bottle. The quality department initiated sporadic tests of bottle sediment. Though the issue was resolved by changing the filter medium, the tests continued, costing several thousands of dollars in labor and lost product every year. An audit may have determined the test did not benefit the brewery any longer.

A brewery that conducts audits well re-engages its employees on their quality values and helps reinforce how the values are articulated in their daily duties.

Audits give the brewery a chance to pause, reflect on current policy, and re-engage the workforce. Brewery managers can look at the entire body of work being done in a quality system, recognize what has changed over the course of a year or several years, and determine if their quality values and strategy are still congruent with what is occurring in the brewery. Though this seems like a broad objective, it actually has real value in terms of employee morale and effectiveness. A quality department that becomes stale, or does not evolve with the brewery, runs the risk of employees being apathetic to the quality system. A brewery that conducts audits well re-engages its employees on their quality values and helps reinforce how the values are articulated in their daily duties. The audit presents an opportunity to keep quality values fresh in the minds of the entire brewery staff. This pays dividends when those values have to be called upon for decisions between quality and production, for example.

CONDUCTING A QUALITY SYSTEM AUDIT

The quality system entails several sub categories, or assessments, and may be performed by the brewery management team, a quality manager, or a third-party consultant. The steps for management to bring a quality system audit into the brewery are:
1. Agree and communicate the importance of the audit to employees and other managers.
2. Decide who will conduct the audit.
3. Determine what areas will be assessed (brewery/cellar, utilities, packaging, shipping and receiving processes).
4. Determine the frequency for conducting the audit(s).
5. Agree on the format and output of the audit as well as who will take responsibility for the output. Summarize what cost savings the audit revealed and what process or procedural gaps the brewery had, and provide a prioritized timeline of actions that need to be taken.

After the brewery has made the decision to conduct audits, and leadership has communicated the need to the rest of the management team and brewery personnel, the next thing to decide is who will conduct the audit and what level of detail the system audit will cover. A third-party auditor is a good way to start if the brewery is new to auditing or does not have the expertise in-house. Sometimes, a newly hired management-level employee in quality is a good person to spearhead an audit, assuming they have the background and have participated in audits before.

The benefits of a newly hired person or a third-party consultant conducting the audit are many. If auditing is new to the brewery, the brewery management can learn audit criteria in a general sense, and then begin to modify for their needs. For example, sometimes a third-party auditor may score an audit, or provide an assessment of a process that is not of primary concern to the brewery. The good news is, if a third-party conducts the audit, the brewery can take or leave the advice without an employee getting hurt feelings in the process. Additionally, the brewery can modify the audit criteria for its own needs the next time around.

For example, an auditor may be stringent on the requirements of malt receiving for silo malt. If they see an unlocked, open hose connection, even though the connection is in a secured area of the brewery campus, they may inform the brewery they require a locked secure cover at the hose connection and mark them down on the audit. The brewery will know that while there is a gap in process of silo security, they have more gaps centered around securing receiving materials. Perhaps the audit made them look more closely at the process of receiving for bag malt. In this case, the brewery team used the audit to influence and improve many processes on receiving security, not just the one the auditor commented on.

> Breaking the system audit down into the sub-components of process assessment (processes), department assessment (policy and people), and product assessment can help narrow the focus.

A third-party auditor allows for this type of learning. After the brewery has determined who will conduct the audit, the next step requires the brewery to be selective on what it will assess. Brewery management must determine what process, procedures, and policy will be reviewed and against what criteria. It is possible all processes, procedures, and policy can be reviewed at one time; however, this is a large undertaking. Therefore, breaking the system audit down into the sub-components of process assessment (processes), department assessment (policy and people), and product assessment can help narrow the focus.

The team responsible for the process in the first place, such as the laboratory, packaging, and the brewing or utilities, usually performs process assessments.

The results for a process assessment can be broken down into one of three categories:
- SOPs exist and are followed. No gaps in process.
- SOPs are mostly in place, gaps in process consistency.
- Significant lack of SOPs or process inconsistency among staff.

The following table shows many different types of process assessments to review. These can be broken into sections and conducted throughout the year. Procedure aberrations that are noted to improve the process should have updated SOPs post-audit. It is not recommended to conduct a full process assessment every year. Processes take time to evolve and change, and it takes time to repair processes in a working operation. About six months after the assessment, a follow-up is recommended on the progress of gap corrections that were found. An example of a quality system audit that is broad and serves a smaller brewery is found in Appendix J.

THE QUALITY SYSTEM AUDIT IN THREE PARTS
Process Assessment

Operation Area	What to Review
Quality control – lab SOPs	Confirm all tests conducted by the lab have a written procedure and that the procedures are followed. If any procedure aberrations are noted, determine if they are critical to fix or if they are an improvement. Check for consistency between staff against a standard.
Taste panel SOPs	Review taste panel training records and taste panel effectiveness. Ensure an SOP exists for all taste panel tests and that the criteria for holding or releasing beer based on panel feedback are clear. Check for consistency between tasters using a test standard.
Brewing quality control SOPs	Confirm all QC tests relevant to the brewing operation (pH, Plato, cell counts, etc.) have an SOP and that it is being followed. If any procedure aberrations are noted, determine if they are critical to fix, or if they are an improvement. Check for consistency between staff.
Brewery process controls	Confirm written process controls and procedures for when out of control conditions exist (e.g., mash-in temperatures, rate of flow for steam, etc.). Review performance of controls (when an out of specification condition occurred, what happened?). Review training effectiveness against SOPs. Review documentation requirements and completeness.

PULLING IT ALL TOGETHER – ASSESSMENT TIME

Cellar quality control SOPs	Confirm all QC tests relevant to the cellar operation (pH, Plato, cell counts, etc.) have an SOP and that it is being followed. If any procedure aberrations are noted, determine if they are critical to fix, or if they are an improvement. Check for consistency between staff.
Cellar process controls	Confirm written process controls and procedures for when out of control conditions exist (e.g., cellar fermentation tanks control settings, pressures, CO_2 recovery, etc.). Review performance of controls (when an out of specification condition occurred, what happened?). Review training effectiveness against SOPs (are employees doing what is expected?). Review documentation requirements and completeness.
Cleaning and sanitation SOPs	Confirm a master cleaning and sanitation plan exists. Confirm cleaning and sanitation SOPs exist and whether they are being followed and documented where needed. What procedures are missing? Do verification steps exist? Are corrective actions clear if a cleaning process failed? This assessment can occur along with a cellar or brewery process control assessment.
Good manufacturing practices	This assessment can be very comprehensive to all GMPs, food safety, and cleaning and sanitation. Usually conducted by a third party. This assessment reviews general GMPs in addition to sanitation and food safety: condition of brewery, visitor policy, pest control, safety and security of the brewery, employee policy, procedures on hand-washing, etc.
Packaging quality control SOPs	Confirm that all QC tests relevant to the packaging operation have an SOP and the SOP is being followed. This includes finished goods checks such as visual inspection of labels, crowns, canning seams, dissolved oxygen, CO_2, etc. If any procedure aberrations are noted, determine if they are critical to fix, or if they are an improvement to current SOP. Check for consistency between staff.
Packaging process controls	Confirm written process controls and procedures for when out of control conditions exist (for example, filler pressures, temperatures, vacuum, labeler glue temperatures, pasteurizer settings, and water treatment). Review performance of controls (when an out of specification condition occurred, what happened?). Review training effectiveness against SOPs (are employees doing what is expected?). Review documentation requirements, and completeness.
Utility quality control SOPs	Confirm that all QC tests relevant to the utility operation (steam water quality, CO_2 purity, and general water quality) have an SOP and the SOP is being followed. If any procedure aberrations are noted, determine if they are critical to fix, or if they are an improvement to the current SOP. Check for consistency between staff.
Utility process controls	Confirm written process controls and procedures exist for out of control conditions (boiler controls for steam delivery). Review performance of controls (when an out of specification condition occurred, what happened?). Review training effectiveness against SOPs (are employees doing what is expected?). Review documentation requirements and completeness.
Quality assurance – brewery procedures	Confirm written procedures exist for maintenance, calibration, and documentation and that the frequency, specifically related to measurement equipment, is being followed. If a brewery has in-line measurement equipment (such as Coriolis flow meters for density, dissolved oxygen meters, alcohol meters, CO_2 meters, etc.) this audit is critical to a quality system functioning as expected. This audit also includes off-line equipment such as scales. In-line and off-line equipment maintenance and calibration may be best performed by the original equipment manufacturer or a certified calibration company. Look for the use of outside certified standard weights, etc., from the National Institute of Standards and Technology (NIST) or similar agency.
Quality assurance – lab procedures	Ensures lab procedures for maintenance, calibration, and documentation exist and the frequency of checks is being followed. Look for the use of outside certified standard weights, etc., from the National Institute of Standards and Technology (NIST) or similar agency.

QUALITY MANAGEMENT ESSENTIAL PLANNING FOR BREWERIES

Food safety controls (HACCP)	Conducted as part of HACCP or other food safety program, a food safety assessment primarily reviews risk assessment (FMEA or other), raw material receiving and logging, traceability of raw materials by lot, trace-back of finished goods to the distributor, glass control, and hazard controls (chemical, physical, microbiological). These are normally conducted by a third party.
Supporting systems	Review training, HR support, and maintenance procedures for completeness as part of an overall process audit. Do these teams support the quality processes?

Figure 8.1: Process Assessment Types

Department Assessment

The department assessment component to a quality system audit provides details to leadership about whether the quality values are still congruent to current practices. The focus is on the quality department.

Operation	**What to Review**
Quality written policy	Does the policy exist and is there evidence it is followed?
Quality training	What training procedures exist? Is there evidence these procedures are followed?
Quality certification of staff	Are employees required to display required skills before they are "signed-off" as a certified quality operator? Are there skill gaps in staff conducting the tests?
Quality responsibilities	Who is doing what related to the quality measurement plan? Is there room for advancing some measurement responsibilities to operations versus the laboratory staff?

Figure 8.2: Quality Department Assessment

Product Assessment

A product assessment should happen every year. Figure 8.3 reviews the components of the product assessment.

Operation	**What to Review**
Quality specifications	Review current and new products. Determine if additional risks are being addressed by the current specifications. Determine if there are gaps in finished goods or raw material specifications.
Product performance	Review consumer data, complaints, and field reports on product performance. Is any particular product, package, or territory performing as expected, better than average, or worse than expected? Territorial differences in consumer complaints may play a role in this data analysis.
Operations performance	Determine if specification or process needs to be adjusted by checking the process capability versus the specification limits in any one product or process that continues to show out of spec or close to spec limits.
Raw materials conformance to specifications	Review supplier performance and conformance to their specifications.

Figure 8.3: Product Assessment Components

The Quality System Audit can be performed in its entirety on a two-year calendar. As start-ups can have many changes to SOPs in the first few years, it is not recommended to develop a full-scale auditing program until the brewery has been in business at least two to three years. Annual product assessments should begin after two years of being in business. If the brewery is still in early stages of developing SOPs, the process assessment early on can help determine what SOP should take priority, and develop a plan to write SOPs as part of a quality program.

The quality audit can show how a brewery has evolved and help guide where it needs to go in the next three to five years. It is a good exercise to routinely conduct a complete round of product and process assessments so the brewery is in a constant state of evolution and quality improvement.

KEY TAKEAWAYS

1000–15,000 BBL Brewery	15,000–150,000 BBL Brewery	150,000+ BBL Brewery
• Quality audits are not in place yet.	• Quality audit shows where gaps exist in a system. It may take 2–3 years to perform another audit.	• Product and process audits are routine.

APPENDIX A

SMALL BREWERY QUALITY MANUAL EXAMPLE

Every brewery should have a foundational quality policy and procedure manual. The manual needs to be referenced frequently and should be updated on a routine basis, every two to three years. This is an example of a small brewery quality manual. There are several components that are placed under separate cover in this manual example, including the quality control plan and finished product specifications. Appendix B has examples of these. Separation may be necessary because these sections of a quality manual are bulky, may only be kept in electronic format, or may be added to frequently. A large brewery may have every item in this manual as a separate quality standard, and include more detail with every standard.

As with every document produced that explains procedures, policies or standards, or templates created to log data, there should be document control or a standard way to number and route approvals. This example shows document control on the first page.

QUALITY MANAGEMENT ESSENTIAL PLANNING FOR BREWERIES

Brewery ABC Quality Manual		
Prepared by: Jane Doe, Quality Manager	Reviewed and approved by: Susan Simpson, Brewmaster	Version: 1.13
Date: 1/13/15	Supersedes: Version 1.12	Date Effective: 2/1/15

Table of Contents

- Quality Policy and Governance
- Good Manufacturing Practices Standard and Pest Control – Under Separate Cover
- Master Sanitation Plan
- Risk Assessment and Control Points
- Glass Policy
- Quality Control and Assurance Plan – Under Separate Cover
- Finished Product Specifications – Under Separate Cover
- Raw Materials Handling Standard – Under Separate Cover
- Recall, Traceability, and Consumer Complaints Policy
- Good Laboratory Practices

QUALITY POLICY AND GOVERNANCE

Quality Policy

It is our Brewery's policy to maintain the highest quality in our products via our brewery operations team, packaging operations team, and quality team. It is our vision to be seen as a world-class brewery and to continuously improve our processes and systems to build confidence in our customers and grow our brand.

Quality Governance (roles and responsibilities):
Quality is directed and governed by the brewmaster and executed by the production manager, brewery manager, and quality manager. A true-to-type or technical brand footprint is maintained by the quality manager and informed by the brewmaster. The quality manager also maintains the quality control and quality assurance plans, risk assessments, and testing procedures. Quality specifications of new products are created during the new product development process by the quality manager and signed off on by the brewmaster. The production and brewery managers must execute their processes so they meet product specifications and quality standards. If production must interpret a specification, perform a corrective action that is non-standard, or otherwise make a quality interpretation of a product, the personnel must have the concurrence of the quality manager on the action. The quality manager is empowered by the brewmaster to make quality-minded, customer-centric interpretations of specifications or corrective actions, and determine if products need further evaluation prior to release (hold and release). All quality deviations are maintained in a deviation log by both the production and brewery managers. All deviations and corrective actions are reviewed with the CEO on a routine basis (quarterly) along with consumer complaints. It is our company's primary quality strategy to continuously improve our operation. We envision our brewery will have self-managed and well-trained teams that effectively manage the quality of their process and outputs.

Org Chart

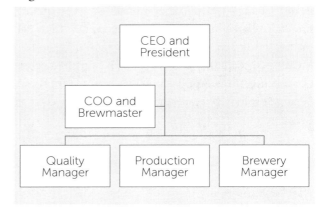

If there is disagreement on any policy, specification, or quality check procedure, the quality manager may escalate to the brewmaster for resolution. If gaps in policy are resolved with the production and brewery managers, the quality manager must bring these to the attention of the brewmaster for final concurrence.

GOOD MANUFACTURING PRACTICES STANDARD AND PEST CONTROL

Under Separate Cover (see Appendix H)

MASTER SANITATION SCHEDULE (MSS)

The Master Sanitation Schedule is developed and maintained by the production manager as part of the asset care procedures. This schedule includes all facility sanitation elements, including pest control, weed/landscape maintenance, spare parts stores, walls, floors, tank exteriors, overheads, and other relevant items. It states what procedure is done, and at what frequency. If a formal SOP exists, the documentation will be linked to the master sanitation schedule. The master sanitation schedule is under a separate cover and is updated frequently.

Sanitation/CIP Policy

The brewery is to be kept in working order, and under strict cleanliness rules as dictated by our GMP policy. Clean-in-place-systems (CIP) are routinely conducted with the following requirements:

Brewhouse Vessels: Brewhouse vessels are cleaned at the end of the brewing week with caustic cleaner as stipulated in the brewhouse CIP SOP. The malt mill is dry-cleaned weekly, according to SOP. Mill and brewhouse cleaning is verified by a supervisor prior to start-up the following week.

Cellar Vessels: All cellar fermentation vessels are cleaned using a caustic wash immediately after emptying, and sanitized with a peracetic wash prior to filling. Fermentation tanks left empty for longer than 12 hours must be re-cleaned and sanitized (not sanitized only). Complete breakdown of the valves, gaskets, etc. is part of the CIP process. Inspection, both visual and with ATP swabs, is required with every CIP.

Bright Beer Tanks: Bright beer tanks are rinsed with phosphoric acid after emptying, and sanitized with peracetic acid prior to filling. Bright beer tanks are fully cleaned using the CIP system (emptied of CO_2, caustic washed, rinsed, and re-sanitized) on a quarterly basis.

Packaging Equipment: Packaging equipment is fully cleaned using the CIP system with rinse, caustic, rinse, and acid washed after a run. Phosphoric acid is left in the lines for cleanliness. If the equipment is left empty for longer than 24 hours, equipment is rinsed and fully cleaned with caustic again. Equipment is fully sanitized in place prior to any run. Conveyers and the filler exterior are kept clean with foaming cleaners at end of production day. The full breakdown and extensive cleaning schedule is maintained by the asset care or maintenance team.

Documentation

All CIP cleanings are documented on the maintenance log for the equipment, production log in the packaging hall, brewhouse log in the brewhouse, and the individual tank records in the cellar. All inspections, sanitizer titrations, and swab verifications are recorded on the respective log or production record. The microbiology sampling plan is coordinated in conjunction with sanitation procedures. Any high counts or areas of concern are investigated and corrective action is taken and documented by the quality control team.

Maintenance of CIP Equipment

CIP equipment (pumps, vessels, spray balls) is maintained as process equipment by the asset care or maintenance team. Inspection and preventive maintenance plans are the responsibility of the maintenance department.

CIP Responsibility

All CIP cleanings are conducted by the cellar, production, or brewing staff. Process verifications of chemical levels, rinse effectiveness, and cleaning validation are also performed by the operations team. In the case of a failed CIP cleaning, as shown by a verification procedure (swab or visual), the immediate corrective action in the form of a re-clean, rinse, or other is the responsibility of operations. All documentation is the responsibility of operations teams.

General Procedures (for detailed procedures, refer to brewery SOPs)
Caustic Cleaning – 2% Caustic
    Recirculated for 20 minutes at 125°F
    Rinsed for 10 minutes and tested for residual (pH) target 7.0
    Monitored by pH strips
    Recorded on CIP log for each tank, production log for packaging line
    Verified by ATP test and microbiological sampling program
Acid Cleaning – 1% Phosphoric Acid
    Recirculated for 20 minutes at 125°F
    Rinsed for at least 10 minutes
    Monitored by pH strips
    Recorded on CIP log for each tank, production log for the packaging line
    Verified by ATP test and microbiological sampling program
Sanitation – Peracetic Acid 200 ppm
    Recirculated for 20 minutes
    Levels checked by titration kit
    Recorded on CIP log for each tank, production log for the packaging line
    Verified by ATP test and microbiological sampling program

RISK ASSESSMENT AND CONTROL POINTS

Process Step	Microbiological/Chemical/Physical Hazards	Critical Control Points
Receiving Raw Materials	Microbiological	No
Water Filtration	Microbiological	No
Milling Malt	Physical (Metal)	No
Mashing	Chemical (Sanitation Residual)	No
Lautering	Chemical (Sanitation Residual)	No
Spent Grain Removal	None	
Boiling	None	
Hop Additions	None	
Fermenting	Chemical (Sanitation Residual)	No
Centrifugation and Filtrations	Chemical (Sanitation Residual)	No
Bright Beer Storage	Chemical (Sanitation Residual)	No
Packaging	Physical (Glass) Chemical (Sanitation Residual)	No
Palletizing and Distribution	Physical (Glass)	No

A full HACCP plan has not been developed for this facility. The risk assessment conducted above highlights the critical risks that are all addressed with GMP, housekeeping, or sanitation policy and procedures.

GLASS POLICY

Scant glass breakage outside of packaging occurs in this operation. Therefore, controls for glass are based on standard housekeeping, prevention, and inspections. Additionally, glass control in the facility is part of GMP policy. On the outside chance that glass breaks in a production area, a record is kept, the corrective action is logged, and verification of clean-up by inspection ensues.

Packaging glass breakage can occur due to poor glass quality, damage during transit, or other rough handling on the line. All glass breaks are documented on the packaging glass record. Production must stop for a full glass clean-up if breakage occurs outside the filler near open containers. The filler has automated spray-down function that performs a full flush of the station and adjacent stations for three rotations. If glass breakage occurs to more than three bottles/100K total, then the filler is stopped, rinsed down, and fully inspected for glass evacuation of housing bells, seals, etc. The packaging manager is responsible for documenting glass breakage in the filler, and inspection of the filler if triggered by this action limit as stated in the quality control plan (QCP).

Glass Breakage Log   Version 2.13   7/13/2015

Date	Time	Location	Type of Glass	Operator	Corrective Action	Verified by	Date	Time

QUALITY CONTROL AND ASSURANCE PLAN – UNDER SEPARATE COVER (SEE APPENDIX B)

FINISHED PRODUCT SPECIFICATIONS – UNDER SEPARATE COVER (SEE APPENDIX B)

RECALL, TRACEABILITY, AND CONSUMER COMPLAINT LOGS

Recall or Product Pullback Documentation and Procedures:

In case of a RECALL, the staff member designated as a recall coordinator and their team will complete the following.

Our recall coordinator is (insert employee name and title):
1. Phone the State Food and Drug Administration (number here) and let them know that we must recall the products involved:
    a. Identity of product(s)_____
    b. Lot number(s) or date code(s)_____
    c. Other information that may help identify the affected product(s)_____
    d. Information on where the product was shipped or sold attached (list of distributors)
    e. Number of cases that were in the affected batch_____
    f. Number of cases that you still have in your control_____
2. If advised to do so by the FDA, we will log the event on the FDA Reportable Food Registry.
3. Take steps to separate and quarantine any of the affected product that is still in our control. Determine whether a corrective action can be taken.
    a. Indicate the action taken here:_____
    b. _____
    c. If a process authority was contacted:
        i. Process authority contacted_____
        ii. Recommendation from process authority (attach letter)_____
4. Contact all distributors on list affected (sales or VP liaison).
5. Contact the FDA if product entered interstate commerce.
6. All phone, in-person, and/or e-mail contacts involved will be logged below.

Notes:_____
_____
_____
_____
_____
_____
_____

In case of a PRODUCT PULLBACK the recall coordinator and team will complete the following.

Our recall coordinator is (*insert employee name and title*)
1. Identity of product(s)_____
2. Lot number(s) or date code(s)_____
3. Other information that may help identify the affected product(s)_____
4. Information on where the product was shipped or sold attached (list of distributors)
5. Number of cases that were in the affected batch_____
6. Number of cases that you still have in your control_____
7. Take steps to separate and quarantine any of the affected product that is still in our control. Under direction of a process authority, determine whether a corrective action can be taken. Indicate the action taken here: _____
8. Contact all distributors on affected list (sales or VP liaison).
9. Record successfully pulled-back product and document the refund to distributors.

Traceability Records

All brewing raw materials, malt, salts, and hops are logged into dock inventory at receipt. Brewing raw materials are recorded on brewing logs. A trace-back audit is conducted once a year from raw material through to all production runs of a small lot of hops. Any other raw materials logged on miscellaneous raw materials log (spices, fruit, etc.). Packaging raw materials are entered into the receiving log. The packaging lots (glass) are captured on the production log for each day.

Product Pullback Audit Form

Date: _____ Time Begin: _____ Time End: _____
Total Time: _____ Raw Material to Be Traced: _____
RAW MATERIAL RECOVERY:
Lot Number(s): _____
Date of Receipt: _____
Amount of Raw Material Involved: _____
Amount of Raw Material Accounted For:
    On Hand: _____ Used in Finished Goods: _____
    Percent of Raw Material Accounted For: _____
Finished Goods Recovery: _____
Code Dates: _____
    Amount of Finished Goods Involved: _____
    Amount of Finished Goods On Hand: _____
    Amount of Finished Goods Shipped: _____
    Percent of Finished Goods Accounted For: _____
Corrective Actions:_____
_____

Signature of Brewery Manager/Team Leader: _____

Consumer Complaints

Consumer complaints are logged on record by the brewmaster in our Excel database, and root cause is investigated by the team. If root cause is identified, corrective actions are noted, documented at management meetings, and incorporated into SOPs. The record is reviewed by senior management on a quarterly basis. If there are returns, a financial record is maintained by the accounting department.

Date	Consumer Name (if provided)	Consumer Address (if provided for refund only)	Product	Date Code	Complaint

GOOD LABORATORY PRACTICES

The lab is a common workspace. There two areas designated by zones for lab work vs. office work. Lab work is conducted at the alcolyzer and bench top that holds the alcolyzer. Center bench is used for taste panel and blending. Office bench is used for paperwork and data entry.

Eating is not allowed in the laboratory on the lab bench (office bench is acceptable). Food is not allowed to be stored in the refrigerator so as not to attract pests. The lab floor, cabinets, and sink area are cleaned by an outside cleaning service. The refrigerator interior and cabinets are maintained by the laboratory employee.

Good laboratory practices are used in the lab and include:
- Lab glassware is maintained in good shape and frequently inspected for cracks.
- Lab glassware is not used for taste panel (to drink directly out of). All blends are poured into proper taste panel glasses.
- Taste panel glass is not used for chemical analysis or collecting samples from the cellar.
- When possible, plastic containers are used to transport cellar samples to the lab (minimal use of glassware outside of the lab).
- Mouth pipetting is not allowed.
- Material Safety Data Sheets (MSDS) for any lab chemicals are kept in the lab and are accessible to all employees. All lab-trained employees must read the MSDS manual yearly.
- No food is allowed in the lab refrigerator.

APPENDIX B

QUALITY CONTROL AND ASSURANCE PLANS

Because within the brewing process there are many points of data to measure, making sense of it all and maintaining some semblance of product quality control requires a simple set of documents explaining this complex system of measurements. In this appendix, there are examples of a Quality Control Plan (QCP), a Quality Assurance Plan (QA), a finished beer specification, and a portion of a process control plan. These standards can become elaborate and overlap one another. This makes updating the plans very labor intensive. To prevent that from occurring, it is good to stipulate the various differences of these standard plans, and avoid repetition of data.

QCP – This plan details the quality checks (QC) conducted on raw materials, in-process beer, or finished beer. These checks all have actionable limits to hold, deviate, blend, or dump batches to prevent out-of-specification beer.

QAP – This plan details the QA checks conducted to ensure proper maintenance and calibration of the measurement system.

Process Standard or Process Control Plan – These standards or plans make sure the process controls are set properly and are running "in control." Actions are taken only if process is so far out of control it will impact a QC check. The process standards can include checks on items such as temperatures, flow rates of steam, CO_2 collection, dosing automation, timing, gassing, pressures, etc. These checks are made while beer making is "in process" and ensure the process is ready or running as normal. If the data is non-discreet, it can be used to improve a brewery's ability to maintain product control, and therefore reduce quality checks. For example, if mash mixer heating cycles maintain precise ramp-up controls, mixing, and mash in, iodine checks may be reduced. Monitoring data from process controls, either manual or automatic, becomes a large part of the quality system. The example used here is for the process of fermentation.

Finished Beer Specification – Brewery size will stipulate how this document is represented. A small brewery can place all specifications related to one beer and brand on one document, and include a visual of labels, packaging, etc. Some larger breweries utilize different blends of finished beer streams to make up a brand. They also may have many types of configurations of packaging to run. Therefore, larger breweries may detail all chemical and physical specifications on a table for finished beer streams, then cross reference the beer streams or blends with packaging requirements (depending on package type configuration) for a specific brand/package SKU.

Brewery ABC Quality Control Plan (QCP)		
Prepared by: Jane Doe, Quality Manager	Reviewed and approved by: Susan Simpson, Brewmaster	Version: 1.13
Date: 1/13/15	Supersedes: Version 1.12	Date Effective: 2/1/15

MICROBIOLOGY SAMPLING PROGRAM

SAMPLES TAKEN				Target Organism	Control Limits		Corrective Action
Sample Point	Frequency	Amount	Media		UTL	Spec	
Wort	N=1/turn	1 mL	UBA	Lactic acid bacteria Wort spoilers Yeast and wild yeast	<1 cfu/mL	10	If above UTL, investigate cleaning procedures. If above spec, hold tank and monitor through fermentation. Do not blend tank until it passes taste panel.
Fermenters (24–48 hrs.)	N = 1/tank	1mL 1mL	UBA+A	Lactic acid bacteria Wort spoilers or water borne organisms	<1 cfu/mL 10	100	If above UTL, investigate cleaning procedures. If above spec, hold tank and monitor through fermentation. Do not blend tank until it passes taste panel.
Pitching yeast (for wild yeast)	1/month	10^6 cells	Lin's media Lysine media Copper sulfate	Wild Yeast	<1 cfu/mL	100	If above UTL, monitor yeast and beer closely. Replace yeast asap. If above spec, yeast must be replaced immediately and affected tanks monitored for wild yeast contamination, off-flavors. Do not blend tanks.

APPENDIX B: QUALITY CONTROL AND ASSURANCE PLANS

Bright tanks	N = 1/tank after fill and N = 1/every 2 weeks in storage.	1 mL	SDA anaerobic	Lactic acid bacteria Other	<1 cfu/mL 10	10 100	If above UTL, audit CIP process, back flush lines. If above spec, hold tank, evaluate for VDK and taste. Do not blend. If pass, release as kegged beer only.
Packaged beer	N = 6/lot sampled randomly in run, C = 0 (No positive samples allowed)	Package contents on membrane 100 mL on membrane filter	SDA anaerobic/ UBA aerobic	Lactic acid bacteria Water borne organisms	<1 cfu/mL	5	If above UTL, audit CIP process, back flush lines. If above spec, hold product for 10 days, evaluate for VDK and re-taste. Release to fast moving market.
Rinse water (bottles, hoses, kegs, etc.)	N = 3/week random hoses, kegs and rinse water	100 mL on membrane	PCA		<1 cfu/mL	10	If above spec, "indicator" organisms, inspect hoses, lines etc., audit process. Replace worn parts. Back flush where possible. Re-sanitize any lines.
Cleaned surfaces (tanks, filler, etc.)	N = 3 swabs/CIP N = 5 swabs at filler startup	100 cm² swab (10x10)	ATP			150	If above spec, re-CIP affected area.
Cold liquor, carbon towers, jetter water	N = 1/quarter	100 mL on membrane	PCA	Water borne organisms	<1 cfu/mL	10	If above spec, review flushing procedures. Re-sanitize tanks and equipment. Replace carbon.
Pitching yeast viability Count/consistency	Every pitch	normally 1/100 dilution required to count	Methylene blue Hemocytometer	Not applicable	≥95% viable	≥90% viable	Monitor affected tanks if above spec for lagging fermentations, off-flavors, etc. Replace yeast ASAP.

Lactic acid bacteria	Gram pos, catalase neg
Wort spoilers or some water borne bacteria	Gram neg, catalase pos

UTL = Upper Tolerance Limit. This limit is usually lower than the specification and allows time for reaction.
Spec = Surpassing the specification requires a more stringent correction.
SDA = Schwarz Differential Agar
PCA = Plate Count Agar
UBA+A = Universal Beer Agar Plus Actidione
ATP = Adenosine Triphosphate Swabs
pos = Positive
neg = Negative
cat = Catalase

QUALITY MANAGEMENT ESSENTIAL PLANNING FOR BREWERIES

ANALYTICAL AND SENSORY SAMPLING PROGRAM

	SAMPLES TAKEN			Limits		Corrective Action
Sample point	Frequency	Amount	Test	LCL	UCL	
Liquor water	N = 1/week	100 mL	pH		8.2	Monitor liquor for excess alkalinity. Determine if additional adjustment in mash is needed based on these results. Monitor tank for salts buildup. Communicate maintenance needs. If any pH test exceeds 8.2, pull duplicate sample and confirm. Conduct root cause analysis and correct. Do not allow longer than 1 week of high pH tests to continue without root cause analysis.
Mash/ sparge water	N = 1 /week	100 mL	pH		8.2	As above.
Mash/ Lauter	N = 1/at end of mash off	1 mL	Starch conversion		Negative	Do not process the mash forward until starch returns negative. This may be after *vorlauf* at 165°F or higher. If conversion is still not completed after *vorlauf*, continue to hold and re-*vorlauf* until converted. Check iodine solution and positive control.
Kettle	N = 1/boil at end of boil		pH Plato	5.2	5.5 < or > 0.5 target	If pH does not meet the desired target, review water chemistry, salt charges, and acid additions. Adjust. If Plato is above or below the UCL, adjust grist weight for successive batches if blending wort. Watch tank.
Fermenters (24-48 hrs.)	N= 1/tank at 24 hours	100 mL min	Plato Aroma pH	Beer type dependent	2.0 delta from wort pH	If beer does not hit target process specification during fermentation, put a quality watch on the tank and prepare for kraeusening if needed. Monitor subsequent fermentations if pitching yeast must be used. If beer over exceeds expected Plato after 24 hours, review pitching procedures and put a QC watch on tank. Plan for blending as needed. Should smell slightly green appley, worty and yeasty. Expect a drop in pH not to exceed 2.0 from wort pH. If it exceeds this limit, check for excessive pitch, over temp fermenter, or other pitch related issue. Hold tank for blending 25%.

APPENDIX B: QUALITY CONTROL AND ASSURANCE PLANS

Pitching yeast	N = 1 day/brink	100 mL	Aroma Viability	N/A N/A	> 90%	Look for any off-notes (meaty, brothy, sour, or veggie) Should smell like fresh bread, slightly alkaline If viability is < 90%, determine if a secondary source can be used. If no secondary source is available, taste yeast, review fermentation profile from prior batch and plan to put a watch on tank through the first 3 days of fermentation. If fermentation does not progress as expected, kräusen if needed. Hold tank for blending 50% max.
Fermenters prior to crash cool	N = 1/tank		Plato and alcohol Diacetyl pH			All parameters must be defined for each test/beer type.
Primary filtered			Turbidity Alcohol Taste pH			All parameters must be defined for each test/beer type. Corrective action as above. Taste profile: Fresh beer, no off-notes – go/no-go
Bright beer or stored beer	N = 1 /tank after transfer N = 1/week every week storage (Diacetyl after 2 weeks)		Plato Alcohol SO_2 BU Color Yeast counts (if necessary) Turbidity Taste Diacetyl/VDK pH			All parameters must be defined for each test/beer type. If SO_2, alcohol, BUs, Color or more than 2 parameters are not within target, hold tank and plan to blend to target. Taste profile: Fresh beer, no off-notes – go/no-go at bright beer.
Packaged beer	N = 1/start-up (inline) N = 3 bottles start and end of run N = 1/keg run N = 1/beer type/week taste		Alcohol Color pH TPO CO2 Taste	Btl CO_2 2.35 Keg CO_2 2.3	TPO 100 ppb Btl CO_2 2.65 Keg CO_2 2.5	All parameters must be defined for each test/beer type. If alcohol, color, or more than 2 parameters are not within target, hold tank and plan to blend to target. Brand specific targets for CO_2 exist. Check brand profile prior to bottling. Taste profile: Full profile, round table. Corrective action only if a major defect is detected by majority of panel. Profile is used for quality monitoring and improvement.

| Water, steam, fob jet | N = weekly samples | | Aroma | | | Aroma – Normal, no off-notes (medicinal, musty). Do not run line if major level of musty or medicinal aromas is detected.

Purge jet and run a maintenance cycle if major defects are detected. |
| CO_2 | N = 1/week or at load in | | Aroma | | | Aroma – Normal, no off-notes (DMS). Reject load if possible. If not possible, purge tank completely prior to next fill. |

RAW MATERIALS AND CARDBOARD PACKAGING SAMPLING PLAN

	SAMPLES TAKEN	Sampling Plan	Corrective Action
Sample Point	Test	UCL	
Malt	Taste malt at receiving. Visually assess for damage. Monitor after milling for mill effectiveness.	Excess flour or excess whole kernels after milling. Off-flavors, mold, or extra moisture.	Report any damage to the supplier. Review loading, handling procedures. Review mill inspection, maintenance.
Hops	Monitor for pest damage, off aromas, lack of vacuum seal.	> 1 bag	Report to supplier. Take representative sample and determine if lot is affected. Decline lot if damage exceeds 10%.
Glass	Monitor for performance.	> 3 defects/100K bottles	Report to supplier. Take representative sample and determine if lot is affected. Decline lot if damage exceeds 10%.
Packaging	Monitor for performance.	> 5% defective	Report to supplier. Take representative sample and determine if lot is affected. Decline lot if damage exceeds 10%.

FINISHED GOODS SAMPLING PLAN

	SAMPLES TAKEN	Sampling plan	Corrective Action
Sample Point	Test	Limits	
All packages – final package	Visual date code, seam or crown integrity, label position, glue bead and placement on board, cleanliness of package, can or dome staining etc.	N = 1/ hour Any major defect or worsening defect halt line and determine corrective action. Hold product back from last good check and perform corrective action sampling.	Check disposal guideline for hold/release.
Bottled products – Finished package	Total Package Oxygen CO_2	N = 3 random bottles at start of run and end of run, c=2 TPO > 100 ppb Limit CO_2--<2.3 >2.85	Monitor data and check for trending. Determine if maintenance is needed if trending near limits, or if one bottle exceeds limit for more than 2 runs. If 2 bottles exceed limit from n=3, hold run and re-sample throughout run. Send high TPO beer to faster moving markets. Check disposal guideline sampling plan for hold/release.
Bottled products crowner	Torques, crown crimp diameter Seal	N = 3 heads every start of run. 100% sampled at the in-line monitor	Monitor for trending, or issues with crowns. If >3 bottles/hour are kicked out, check crimps and seal of discarded bottles. Validate CO_2 levels. Adjust in-line monitor if needed. Conduct root cause analysis at filler.
Bottled products	Fill height	100% sampled at fill check monitor in-line	If >3 bottles/hour are kicked out, check levels of the discarded bottles and validate levels. Adjust fill height monitor if needed. Conduct root cause analysis at filler.

QUALITY ASSURANCE PLAN EXAMPLE

The best way to prepare a brewery QA plan is to review what is measured in the QC plan and determine how the measurement tool will be calibrated, maintained, or checked by a standard. The example below is for pH probes offline and in-line. This is a summary table. Specific operating procedures (SOPs) for the test can be referenced in this table.

Brewery ABC Quality Assurance Plan		
Prepared by: Jane Doe, Quality Manager	Reviewed and approved by: Susan Simpson, Brewmaster	Version: 1.13
Date: 1/13/15	Supersedes: Version 1.12	Date Effective: 2/1/15

Measure	Process points	SOP number	Calibration or standard check frequency	By
pH	Water – in-line	QC SOP_X32	1x/qtr. remove and test in standards. Calibrate according to procedures.	QC Tech
	Mash	QC SOP _ M21	Check to 4 and 7 pH standard every use. Clean probe or replace the electrode solution as slope decays or once every quarter.	Operator QC Tech
	Fermentation	QC SOP _ F10	As above	
	Finished beer	QC SOP _ P12	As above	

APPENDIX B: QUALITY CONTROL AND ASSURANCE PLANS

PROCESS CONTROLS STANDARD EXAMPLE

In this type of quality standard, simply state what parameters are measured, where, at what frequency, and where the data is recorded or documented. Do not get into target numbers in this table. Those are left for individual recipes. Also, this is not the place to document the QC plan, as it is redundant and, more importantly, it is a process output. Instead, focus on the process inputs that are measured at each step of the process. This is done for a couple of reasons. As the brewery improves in controlling the variation of the inputs, the quality outputs (or QC measurements) also become less variable and more controlled. In the example here we show the fermentation process inputs. If the accuracy and precision of the brewery's instrumentation is known, it can be included in this table.

In addition to the process control general standards, as below, each wort stream will have a recipe with the specific requirements for control as well. This recipe may be in the form of a recipe sheet or other template. The controls included in the recipe sheet would be the same as mentioned in a process control standard, but the recipe sheet contains the exact targets for each measure. Recipe sheets also include the timings of additions, weights of raw materials, precise target for times and temperature control, etc. The recipe sheet should be where the brewery records exact process control targets.

Brewery ABC Process Control Standards		
Prepared by: Jane Doe, Quality Manager	Reviewed and approved by: Susan Simpson, Brewmaster	Version: 1.13
Date: 1/13/15	Supersedes: Version 1.12	Date Effective: 2/1/15

Process	**Process Point**	**Monitoring Parameter**	**Frequency**
Fermentation	Fermentation tank fill	Time to fill Empty temperature Full temperature Yeast pitch rate	Document for every fermentation on fermenter log
	Fermentation	Temperature rise CO_2 production normal Yeast dump day, amount	Document for every fermentation on fermenter log
	Fermentation tank crash cool	Temperature decrease rate	Document for every fermentation on fermenter log

FINISHED BEER SPECIFICATION EXAMPLE

	Brewery ABC Finished Beer Specifications	
Prepared by: Jane Doe, Quality Manager	Reviewed and approved by: Susan Simpson, Brewmaster	Version: 1.13
Date: 1/13/15	Supersedes: Version 1.12	Date Effective: 2/1/15

Beer Stream	FG +/- 0.2	ABW +/- 0.2	BU +/- 3	Color SRM +/- 1	VDK/Diacetyl + 10 ppm	Turbidity +/- 25	pH +/- 0.2	SO_2 +/- 0
Beer A	3.2	4.3	34	15	< 30 ppm	< 100	4.3	< 10 ppm
Beer B	3.5	4.5	45	9	< 30 ppm	< 10	4.4	< 10 ppm
Beer C	2.8	3.5	23	5	< 30 ppm	< 10	4.4	< 10 ppm
Beer D	3.8	4.8	50	40	< 50 ppm	Not Measureable	4.5	< 10 ppm

Note: Specification limits are stated as +/- in the column header. Do not release any beer that fails these limits. Control limits can also be calculated statistically for a series of data. If processes are out of control (trending or outside of limits), conduct corrective action as stated in the control plan.

APPENDIX C

HACCP RISK ASSESSMENT AND CRITICAL CONTROL POINTS

This example shows some of the detail that should go into a HACCP risk assessment for a brewery. The physical hazard of glass is likely one of the most critical points to control for most breweries. This shows a process of thinking through the hazard, the types of controls the brewery had in place, and how they would monitor the critical control point.

Food Safety Hazard Analysis Table: For each process, summarize the food safety hazards (chemical, physical, microbiological) that are known to exist for the industry. If there is evidence to support the hazard analysis in published documents, use it here.

Process		Packaging	Step #	19
Food Safety Hazard and Cause		***Likelihood of Occurrence**	***Severity of Adverse Health Affects**	**Control Measures**
		*L = Low M = Medium H = High		
Chemical – Residual sanitizer		L – None	N/A	SOP
Physical – Glass		M – Some glass breakage during filling. Potential for glass penetration into empty containers.	M–H – Glass	Auto spray down for three rotations after glass breakage in filler. Twist rinse prior to filler for any glass that is broken on conveyer into filler.
Microbiological – Pathogen growth		L – Hurdles of alcohol, CO_2 and presence of hops	N/A	Cold fill, aseptic filling, SOPs

Critical Control Point Justification Table: For each hazard identified as likely or severe, fill out a CCP justification. Example below:

Step #	Q1. Does the process step reduce contamination to an acceptable level? If Yes – this step is a CCP If No – move to Q2	Justification for decision	Q2. Could the product become contaminated in excess of acceptable levels or increase to unacceptable levels? If No – this step is not a CCP If Yes – move on to Q3	Justification for decision	Q3. Will a subsequent process step reduce contamination to an acceptable level? If Yes – this step is not a CCP If No – this step is a CCP	Justification for decision	CCP No
19	Yes	Yes – Auto spray down and twist rinsing will reduce this hazard					1

APPENDIX C: HACCP RISK ASSESSMENT AND CRITICAL CONTROL POINTS

Critical Control Point Monitoring Summary Table: For each CCP identified in the justification table, state how the process will be monitored and controlled.

Process Step	CCP No.	Critical Limits	Monitoring				Corrective Actions		
			Procedure	Frequency	Responsibility	Records	Procedure	Responsibility	Records
18	1	> 3 PSI water pressure	Check twist rinse spray pressure. Gauge is clearly identified with CL.	1x/day	Operations	Production record	If lower than 3 PSI, adjust pressure. Monitor throughout day.	Operations	Production record
19	2	Auto spray down after each bottle breakage	Verify auto spray occurred, and at correct stations via brew log records	Every breakage in filler	Operations	Production records	Stop filler, spray down correct valves, inspect for glass penetration on gaskets.	Operations	Production record
19	2	> 3 bottle breaks in 1 hour	Monitor production record for > 3 breaks per hour	Every hour	Operations	Production records	Stop filler and spray down. Check for glass penetration in gaskets. Notify quality to check glass for quality defects and if glass lot needs to be held.	Operations	Production record

APPENDIX D

FAILURE MODES EFFECTS ANALYSIS TABLE EXAMPLE

In this example, we show a Failure Modes Effects Analysis (FMEA) of the fermentation step. This may be helpful during the design phase of adding in new automation or a new cellar. The FMEA is also a good thought exercise for new trainees to illustrate why and when certain checks are conducted. Prior to starting the FMEA, it is important to agree on the numbers representing rating levels for the severity, likelihood, and detection columns. A severity rating of 1 means there is no real detectable quality issue; a severity rating of 9 or above may mean the beer will have to be dumped. A likelihood rating of 1 means it is highly unlikely to occur, while a likelihood rating of 8 or above may mean the failure has happened and is an ongoing issue. A detection rating of 1 indicates a strong ability to detect the failure, while 10 indicates no ability.

The risk priority number (RPN) is a good way to help a team prioritize quality issues. It is computed by multiplying the three rating factors together. In this example, icing in the heat exchanger has been an issue before, causing the beer to be dumped. A check at 24 hours is good, however, it is not immediate. Because the RPN is highest for this failure mode, the team can choose to put a request in for an alarm system for the temperature on the knockout side of the heat exchanger. This will reduce the risk of this failure and bring it in line with the other risks.

Process	Potential Failure Mode	Effects of Failure (if it were to occur)	Severity of Failure	Potential Causes	Likelihood of Occurrence	Current Controls	Ability to Detect the Failure	Risk Priority # (RPN)
Fermentation	Yeast not pitched	Fermentation would not start	9	Incorrect tank hooked up	3	Redundant checks at the tank and at the control panel 24-hour check	1	27 18
				Pump failure during pitch	2	Pump alarm	1	18
	Wort not cooled	Fermentation would not start Yeast would die in tank	9	Icing bridge in heat exchange	5	24-hour check	3	135
				Incorrect temperature setting at heat exchange	1			27
	Tank not sanitized	Microbial contamination	8	Incorrect levels of cleaners or sanitizer	4	Titration	1	24
				Incorrect tank sanitized	2	ATP swab	2	32

APPENDIX E

HACCP PROCESS MAP WITH CCPS

This simplified version of a process map shows how one brewery decided to control food safety hazards in packaging. Note that the level of complexity of this map is not critical. The goal is to quickly and easily explain to an outsider what is considered a critical step and what is done in that step. This map focuses on food safety in packaging, but it could be modified to include all product quality checks. Each process step needs to start with an action verb.

Brewery ABC HACCP Plan: Process Flow Chart Packaging		
Prepared by: Jane Doe, Quality Manager	Reviewed and approved by: Susan Simpson, Brewmaster	Version: 1.13
Date: 1/13/15	Supersedes: Version 1.12	Date Effective: 2/1/15

(Author note: This is a partial process flow chart for packaging. Steps 1–12 would be written separately to support brewing and fermentation steps.)

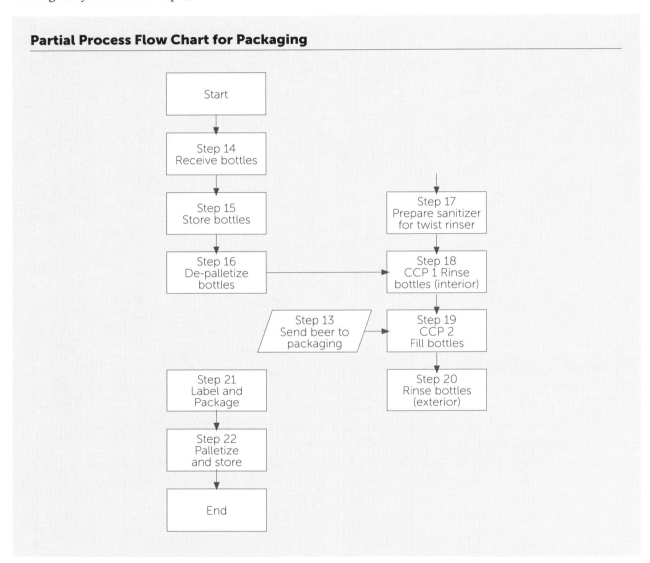

This is a flow chart for a packaging operation. Note that steps 18 and 19 have a Critical Control Point (CCP) associated with them.

APPENDIX F

STANDARD OPERATING PROCEDURE (SOP) EXAMPLE

This is a simple example of a cleaning operation SOP written with the help of a supplier. These can very easily become more detailed, depending on the brewery and the need of the staff to learn directly from the SOP. If the brewery is becoming more complex, make sure a simplified version of an SOP is ready and available for use as a training tool. As the directions become lengthy or more complex, the new trainee can only manage a little information at a time. Start small with an overview and then increase the level of detail to match experience. Some companies choose to write two different levels of SOPs, a short bulleted version of the steps, and the longer detailed version. SOP writing can be tedious and overwhelming. It is best to split the duty between brewery staff, quality team members, management, and one editor to ensure consistency.

Brewery ABC Brewhouse Preparation SOP		
Prepared by: Jane Doe, Quality Manager	Reviewed and approved by: Susan Simpson, Brewmaster	Version: 1.14

Purpose: To prepare kettle prior to start-of-week production. Ensure a clean and sanitary state

Frequency: Weekly post production

Equipment Needed: Secondary container and titration kit

Chemicals:

Chemical	Dispenser	Concentration	Temp	Time
Caustic 50145	Carboy	2.56 oz./gallon	160–180°F	30 min
Oxidizing additive 30232	Carboy	0.64 oz./gallon	160–180°F	30 min
HD acid 95252	Carboy	1.28 oz./gallon	120–140°F	30 min

Safety Information:
Read all Material Safety Data Sheets (MSDS) prior to starting this operation. Perform lock out tag out (LOTO) as stated for this vessel.

Safety Gear

- Chemical resistant gloves (nitrile green, elbow length)
- Safety glasses, face shield for dispensing
- Sanitation/waterproof boots
- Long pants, long sleeves
- Sleeve guards

Procedure:

Step 1: Preparation

- Cover electrical motors and outlets.
- Perform LOTO process.

Step 2: Pre-rinse

- Thoroughly rinse the kettle with the hose removing large hot break and other debris.
- Drain the pre-rinse liquid from the drain valve.

Step 3: Caustic clean

- Add 60 gallons of water to the kettle. Pour 2% caustic solution into kettle (2.56 oz./gallon).
- Shut kettle door and initiate CIP loop.
- Monitor temperature of loop—liquid needs to reach 160°F for 30 minutes.
- Monitor the caustic solution after initial loop, and 20 minutes into CIP with titration kit. Need to maintain a 1.5–2% solution. If in excess of 2.5%, stop process and add more water to reach a proper dilution. If the percentage drops below 1.5%, stop the process and add enough caustic to bump the solution 0.5%.

Step 4: Final rinse

- After wash cycle is complete, dump caustic into retention tank.
- Initiate the rinse cycle and run for 20 minutes at 120–140°F.
- Test last rinse water for pH. Target is 7.0.

Step 5: Return to normal operations mode

- Remove all motor housing covers and LOTO.

Step 6: Monitoring and verification

- Document all data in the sanitation log for the kettle.
- Visually assess kettle cleanliness and document.
- Keep kettle door open to dry.

Modifications:

1. Once every 6 weeks add the 0.5% oxidizing additive to the kettle to remove protein, and run cycle for 60 minutes.
2. Once every trimester run 1% acid clean for 30 minutes at 120°F PRIOR to caustic clean.
3. Do not run acid wash cycle and follow with oxidizer cycle. Keep these on separate weeks of cleaning.

APPENDIX G

QUALITY INSPECTIONS FOR MAINTENANCE

This table summarizes the common quality checks that should be of the most concern for maintenance personnel. The maintenance department should be responsible for monitoring the data related to these checks and being proactive when it comes to equipment. They may have additional checks they perform in order to monitor equipment such as vibration analysis, boiler water chemistry, etc. Details of additional maintenance checks or actions to take because of these checks can be found in other texts on beer packaging and brewery maintenance. This table is a good training table for those involved in maintenance to understand how improperly conducted proactive maintenance affects product quality.

MAINTENANCE QUALITY INSPECTIONS

Equipment	Check	Quality Concerns
Filler – finished goods	Dissolved oxygen CO_2 levels	Gassing levels are critical for taste profile.
	Fill height	Critical for correct tax reporting. If heights vary too much, this can also become a customer complaint issue.
Bottle crowning	Crimp diameter Torque Sonic-check (tap tone or other) for crown seal, bull nose, or other off-centered crowns	The crown seal is a critical quality control point. Any leakage will result in under-gassed beer, and possibly microbiological infection (sour beer) in the field.
Can seaming	Body hook length, cover hook length, C-sink seam height, seam overlap, seam thickness, can buckling, wrinkles	Improper seaming can cause major damage to beer. Any leakage will not only cause one can to be low in CO_2, but also has the potential for microbiological spoilage. Additional damage can occur if the affected can becomes de-gassed enough to wrinkle, and leak, causing potential corrosion of other cans and entire pallets to be damaged by moisture or corrosion.
Labeler	Glue usage, straight application, centered neck and body labels	Excessive glue use may cause labels to "swim" on the bottle and become distorted. Poorly applied labels affect the consumer's brand perception.
Product pumps	Dissolved oxygen (DO)	Worn seals in pumps can cause beer flavor to become damaged by DO. Beer will taste old, possibly "skunky," "sherrylike," or "papery" earlier than anticipated.
Centrifuge	Dissolved oxygen (DO)	As above.
Kettle boil	Flavor (Dimethylsulfide [DMS] or creamed corn), low extract, low alcohol	If steam pressure or kettle jacket is not cleaned and maintained, the result is lower efficiency of the boil, and possible off-flavors.
Lauter tun and mash mixer	Sparge and mash water temperature	If temperatures cannot be held and maintained throughout these two processes, the result is possible low alcohol, lower extract yield, and throughput issues.
Mill	Grind consistency	If mill has gap issues, or rollers are not maintained for sharpness, the result can be lower extract yield, higher use of malt, and flavor implications.

APPENDIX H

GOOD MANUFACTURING PRACTICES (GMP) POLICY EXAMPLE

A GMP policy changes and grows with a brewery as more personnel, risks, and complexities are added to the brewery. However, some basic requirements should be in place no matter what size brewery you are operating. This policy is fundamental to making beer and beverages for the public, and it is required by the FDA to have it in place. A written policy ensures clear and concise guidelines for all brewery personnel to read and follow. A more detailed type of policy can be found on the MBAA website (mbaa.com/brewresources/foodsafety/haccp/Pages/documents.aspx) and, if requested, from your malt supplier.

Brewery ABC GMP Policy		
Prepared by: Jane Doe, Quality Manager	Reviewed and approved by: Susan Simpson, Brewmaster	Version: 1.13
Date: 1/13/15	Supersedes: Version 1.12	Date Effective: 2/1/15

PERSONAL

- Cleanliness: All persons working in direct contact with beer, beer contact surfaces, and beer packaging materials shall conform to the hygienic practices while on duty to protect against contamination of beer. Methods include but are not limited to:
  - Maintaining personal cleanliness.
  - Washing hands thoroughly before starting work, after each absence from the work station, and at any other time when hands may become soiled or contaminated.
  - Removing all unsecure jewelry and any other objects that might fall into beer, equipment, or containers.
  - Maintaining the proper use of gloves. If gloves become soiled from becoming involved in other activities, they must be changed before continuing any work with beer to prevent contamination.
  - Wearing hairnets, beard nets, or any other effective hair restraint as appropriate. Beard covers are required for any open tank operation.
  - Storing clothing or other personal belongings in areas other than where beer is exposed or where equipment or utensils are washed. Wearing clean, cut-free clothes. Covering armpits with short sleeve shirts, minimum.
  - Confining the following to areas other than where beer may be exposed or where equipment or utensils are washed: drinking beer, chewing gum, drinking beverages, or using tobacco.
- Disease control: If a person reporting to work in the brewery is showing signs of illness, open lesions, boils, infected wounds, or shows there is a reasonable possibility of food poisoning, they will be sent home.

FACILITY GROUNDS

- Methods for adequate maintenance of grounds to decrease risk of contamination include but are not limited to:
  - Properly storing equipment, hanging all brooms, brushes, squeegees, etc.
  - Removing litter and waste at the end of each day.
  - Floors and drains need to be properly cleaned and scrubbed at the end of each production day to maintain cleanliness from the floor up.
  - If something can be moved, it will be moved and the floor mopped, instead of mopping around it.
  - Adequately draining areas that may contribute to contamination to beer by seepage or foot-borne filth, or providing a breeding ground for pests and bacteria.
  - Maintain roads, yards, sidewalks, and alleyways so that they do not constitute a source of contamination in areas where beer is exposed. Inspect and take action to exclude pests, dirt, and filth as needed.
  - Provide sufficient space and be constructed in such a manner that the floors, walls, ceiling, and any permanent object in between may be adequately cleaned and kept clean and in good repair.

SANITARY OPERATIONS

- General maintenance fixtures, shelves, and any other physical facilities shall be kept clean and in repair sufficient to prevent beer from becoming contaminated.
- Cleaning and sanitizing of utensils and equipment shall be conducted in a manner that protects against contamination of beer, beer-contact

surfaces, and beer-packaging materials.
- Toxic cleaning compounds, sanitizing agents, and pesticide chemicals shall be identified, held, and stored in a manner that protects against contamination. All beer-grade lubes are labeled, and non-beer-grade lubes are stored separately away from the production area.
- Pest control: No pests should be present in any processing area of the brewery. The use of insecticides or rodenticides is permitted only under precautions and restrictions that will protect against contamination of beer.
- Sanitation of beer-contact surfaces such as tanks and filler shall be cleaned as frequently as necessary to protect against contamination of beer.
  - Non-beer-contact surfaces used in operation of a beer processing brewery should be cleaned and sanitized as frequently as necessary to prevent contamination.
  - Single-service items (such as one-time-use paper cups and paper towels) should be stored in appropriate containers and shall be handled, dispensed, used, and disposed of in a manner that protects against contamination.
- Water supply: The brewery shall be equipped with sanitary facilities and accommodations including but not limited to:
  - The water should be sufficient for the operations intended and shall be derived from an adequate source.
  - Any water that contacts beer or beer-contact surfaces shall be safe and of adequate sanitary quality.
  - Running water at a suitable temperature and under pressure, as needed, shall be provided in all areas where required for the processing of beer, for the cleaning of utensils and beer-packaging materials, or for employee sanitary facilities.
  - Records shall be kept that show "proof of potability" of the water supply. The schedule of testing may vary based on the source of the water.
- Plumbing shall be of adequate size and design and properly installed and maintained to:
  - Carry sufficient quantities of water to required locations throughout the brewery.
  - Properly convey sewage and liquid disposable waste from the brewery.
  - Avoid constituting a source of contamination to beer, water supplies, equipment, or utensils, or creating an unsanitary condition.
  - Adequate floor drainage shall be provided if floors are subject to flooding-type cleaning or where normal operations release or discharge water or any other liquid waste on the floor.
- Toilet facilities shall be provided and maintained. Compliance with this requirement may be accomplished by:
  - Maintaining the facilities in a sanitary condition.
  - Keeping facilities in good repair at all times.
  - Provide self-closing doors that do not open into areas where beer is exposed to airborne contamination, except where alternate means have been taken to protect against such contamination.
- Hand washing facilities shall be adequate, convenient, and furnished with running water at a suitable temperature. Compliance with this requirement may be accomplished by providing:
  - Hand washing and, where appropriate, hand sanitizing facilities at each location in the brewery where good sanitary practices require employees to wash and/or sanitize their hands.
  - Sanitary towel service or suitable drying devices are provided.
  - Proper signage directing employees handling unprotected beer, unprotected beer-packaging materials, or beer-contact

surfaces to wash, and, where appropriate, sanitize their hands before they start work after each absence from post of duty, and when their hands may become soiled or contaminated. These signs may be posted in the processing rooms and in all other areas where employees may handle such beer, materials, or surfaces.
- Trash shall be conveyed, stored, and disposed of to minimize the development of odor, to minimize the potential for the waste becoming an attractant and harborage or breeding place for pests, and to protect against contamination.

EQUIPMENT AND UTENSILS
- All equipment and utensils shall be of a material that can be adequately cleaned and properly maintained. No wood handles or natural fiber brooms allowed. All equipment and beer-contact surfaces shall be made of a non-toxic material and designed to withstand the environment of their intended use.
- Seams on beer-contact surfaces shall be smoothly bonded or maintained so as to minimize accumulation of beer particles, dirt, and organic matter and, in turn, minimize the opportunity for the growth of any microorganisms.
- Equipment that is in the manufacturing area and that does not come into contact with beer shall also be so constructed that it can be kept in a clean condition.
- All manufacturing systems, including pneumatic and automated systems, shall be of a design and construction that enables them to be properly maintained in appropriate sanitary condition.
- Each cold storage compartment used to store and hold beer products capable of supporting growth of microorganisms shall be fitted with an indicating thermometer.
- Instruments and controls used for measuring, regulating, or recording temperatures, pH, degrees Plato, dissolved oxygen, or other conditions that control or prevent the growth of undesirable microorganisms in beer shall be accurate and adequately maintained.
- Compressed air or other gases mechanically introduced into beer or used to clean beer-contact surfaces or equipment shall be treated in such a way that beer is not contaminated with unlawful indirect beer additives.

LIGHTING AND GLASS
- Lighting is an unavoidable source of glass in the facility. Sealed fixtures and/or coated shatterproof bulbs shall be used to protect beer products, equipment, and packaging materials from glass.
- Glass bottles will be monitored for excess breakage. If the limit is exceeded, glass will be pulled off the line and quarantined. Any glass that breaks in the filler will require additional rinse down, monitoring, and filler maintenance.

VISITORS AND SALES PEOPLE
- Ask visitors to sign the visitor registry and to wait in the reception area until they are greeted by the packaging or brewing manager. Provide infrequent and first-time visitors/sales people with a copy of the GMP policy. Ask them to review the policy before being issued a visitor or contractor badge. Affix a sticker to the badge with the date of the visit clearly written on it.

MAINTENANCE AND CONSTRUCTION CONTRACTORS PRODUCT PROTECTION GUIDANCE
- Outsiders coming into the facility must observe basic GMP guidelines to assure product safety. Review the person's credentials and business card or other company identification. All contractors who will be entering the production area will be asked to agree to the GMP policy. Contractors who will be performing significant maintenance, repair, or construction work in the facility shall receive additional instructions, related product protection, and GMP requirements. It is the responsibility of the person contracting the work to review guidelines with

contractors. Contractors who are known to you and who are working in the facility on a routine basis may be allowed access without an escort.
- It is critical that all contractors working in the facility pay careful attention to protecting the integrity of our products. All contractors must agree to comply with the following guidelines, and to instruct all of their personnel to observe the following rules:
  - Each contractor employee performing work in Brewery facilities shall agree to follow the personal practices outlined in the Visitors and Contractors GMP Guidance. The contractor shall review these requirements with all employees prior to their entering the worksite. The contractor shall ensure that new employees receive a copy of the guidance. Workers shall follow these guidelines at all times.
  - Contractors shall assure that work areas are kept clean and free of clutter or debris. Small parts, tools, nuts, bolts, etc. shall be properly stored and not allowed to accumulate in work areas.
  - Doors and other openings to the brewery must be kept closed or screened. When the nature of the work requires that a door or other exterior opening be left open for an extended period of time, the contractor shall notify the area supervisor or brewery quality control before commencing work.
  - Any time work is performed that has the potential to generate contaminates, appropriate action shall be taken to protect product and equipment. Such action includes removing product (including bag or drummed product and materials) from the area, covering product, and covering equipment. Work with the potential to generate contaminates includes welding, cutting, grinding, electrical work, and any work involving small parts.
  - As much as possible, work will not be conducted in active processing or production areas. When such work is unavoidable, the contractor shall discuss the work to be done with the area supervisor to assure that appropriate actions are taken to protect the product.

APPENDIX I

NEW PRODUCT QUALITY CONTROL PLAN EXAMPLE

When a new product is introduced to a brewery, a risk assessment (in the form of HACCP for food safety, FMEA for quality, or some other hybrid assessment) helps the brewery understand what additional quality checks are needed in order to validate if potential or theoretical quality issues do occur or monitor known risks that are being introduced. Once this assessment is completed, the extra risks and controls introduced for the product should be communicated broadly by summarizing using a standard template. This appendix shows what that may look like for introduction of a beer with "wild" yeast to a brewery. Pictures of the product's new label and other photos may be included. This document can be used in production meetings to discuss the beer and the risks, and to engage the operators at all levels to be responsible for new risks.

Brewery ABC New Product Quality Control Plan		
Prepared by: Jane Doe, Quality Manager	Reviewed and approved by: Susan Simpson, Brewmaster	Version: 1.13
Date Reviewed: 1/13/15		Dates Effective: 2/1/15–5/1/15

Product Name: Beer on the Fly
Product SKU: 51-213
Packages: 6-24 bottle, 2-12 bottle, 12-750 mL bottle
Date to Begin Production: 2/1/15
New Product Control Plan Dates Effective: 2/1/15–5/1/15
Beer Description True to Type: Ale made with Saison yeast and 10% barrel aged with naturally present organisms. Tart and slightly yeasty/fruity.
Risks Identified: Microorganisms such as Lactobacillus, Pediococcus, and Brettanomyces may be present in this beer prior to micro-sterile filtration. It will be important to conduct extra monitoring from the blend tank forward to packaging. Process changes include extra cleaning and sanitation times.

Beer Specification

Beer Stream	FG +/- 0.2	ABW +/- 0.2	BU +/- 3	Color SRM +/- 1	VDK/Diacetyl + 10 ppm	Turbidity	pH +/- 0.2	SO_2
51-213	3.2	4.3	25	9	<30 ppm	<50	4.3	<10 ppm

APPENDIX I: NEW PRODUCT QUALITY CONTROL PLAN EXAMPLE

New Product Control Plan

Process Step	Process Control Change	Quality Monitoring Change	Action Limits	Corrective Action
Centrifugation – Filtration	Loop through sterile filter	Turbidity – Every 30 minutes	>50	Stop filtration. Reprocess beer.
		Microbiology checks (SDAA, UBA) at start-up and end of run	Breakthrough of any microorganisms	Review filtration records. Hold beer until sampling and further microbiology checks prove shelf-stability. Dispose of beer to fast-moving market.
Centrifugation – Packaging Sanitation	Clean loop with 2% caustic at 150°F for 25 minutes	None		
Packaging Microbiology Monitoring		Take 3 additional ATP swabs the day after running this beer prior to start-up of filler	>150	Re-sanitize filler
Hold Beer	All finished beer must be held until bright tank microbiology records are reviewed.		See above	

This quality plan is in effect for the first 3 months of production at which time process control will be reviewed.

APPENDIX J

GENERAL AUDIT REPORT

This generic audit report illustrates some of the questions an outside quality systems auditor may ask of brewery management. It can be conducted in-house with a team of managers. Done periodically, an audit can help highlight gaps in the quality system and align staff and the brewery management team on which needs require focused attention. As the brewery grows and expands, this report can guide brewery management on how to allocate resources for quality systems management. The first page of an audit can be used to summarize progress and share with senior leaders.

BREWERY QUALITY SYSTEMS AUDIT

Overall Summary

Product Assessment:
Excellent   V. Good   Fair   Poor

People and Processes to Support Quality:
Excellent   V. Good   Fair   Poor

Quality Control/Assurance, Food Safety, Quality Department:
Excellent   V. Good   Fair   Poor

Maintenance and GMPs/Sanitation:
Excellent   V. Good   Fair   Poor

Conducted on: _____ By: _____

GENERAL INFORMATION

1. What are the primary product quality issues experienced over the past year as analyzed by held beer, consumer complaints, or lost beer?
2. What are the quality program strengths?
3. What are the quality program weaknesses?

PEOPLE AND PROCESSES TO SUPPORT QUALITY SYSTEM

Communication and Training

1. What communications are performed daily, weekly, monthly, yearly?
2. Communication feedback loop. How does this work?
   a. Team members report their production, quality, and safety results to management.
   b. Team members have a feedback loop for how well they are performing.
3. What are the roles and responsibilities for product quality?
   a. There are clear team roles/responsibility for product quality.
   b. Team members know what is expected of them in terms of measuring quality.
   c. We have routine meetings and these are helpful.
   d. There is a quality point person established on every shift.
4. Team structure
   a. We have fully empowered teams; they are organized and responsible for their own work area in quality, safety, cost, sanitation, and maintenance.
   b. Teams understand their roles and how and when to communicate with one another.
5. Communication in general is timely and appropriate for the level of urgency.
   a. Operations teams receive standard brewery goal reports from management.
   b. Only occasionally does production fall out of control in terms of specifications. The focus of teams is on continuous improvement.
   c. The strategic connection to teamwork and projects is apparent.
6. Training and background of staff – general areas/gaps
7. Packaging, brewing, shipping, and utilities processes
   a. SOPs are in place for quality checks
   b. Evidence shows team is following SOPs
   c. Hold/release protocol is in place

Brewery Goals, Strategy, and Support Teams
1. Overall strategy
   a. Brewery's strategy is clear and concise. We have a clear desired mission/vision that has been articulated from the top down.
2. Operations planning and communications are well established to the sales metrics.
3. Human resources
   a. Supports individual goal setting, awareness of culture, team building, and leadership training.
   b. Organizes and supports training of new employees.
4. Operations Strategy
   a. Inventory management
   b. Operations planning

Brewery and Cellar
1. SOPs are written, followed, and clear. Quality checks are conducted by appropriate personnel. Personnel react to check data.

Packaging and Warehouse
1. SOPs are written, followed, and clear. Quality checks are conducted by appropriate personnel. Personnel react to check data.
2. Recall/traceability program is in place and audited frequently.
3. GMP inspections are evident by the cleanliness of the area.
4. First-in and first-out (FIFO) is followed for all raw materials, including labels, cardboard, and six-pack carriers.

QUALITY CONTROL AND ASSURANCE/ QUALITY DEPARTMENT ASSESSMENT
QC Planning
1. Quality manual and specifications:
   a. Written and well communicated.
   b. We have standard and repeatable reactions quality specifications that are out of control.
2. Our suppliers:
   a. Qualified on specific quality performance criteria and service.
   b. Partners in business assisting in continuous improvement.
3. Our inventory management:
   a. Traceability and tracking.
   b. We have a mock recall system.
   c. We have a recall program.
4. Quality control tests:
   a. Control points of the process such as pressures, temperatures, air flow, etc. are well understood (operational controls that impact the finished good specification).
   b. We have a planned frequency of tests. Every test has an SOP and a reaction plan.
   c. Specifications are clear, written, and customer-based.
   d. Specifications are well communicated and easily followed.
   e. When specifications are exceeded, we react at the right level.
   f. We have a routine schedule for calibrations of measurement equipment.
   g. Specifications exist but we have inconsistency in reaching specifications.
   h. Our specifications are customer-derived and prevent overreaction to out-of-specification product.
   i. Variability of data is understood.
   j. We have good test reliability and repeatability.
   k. Data is reviewed and has clear purpose.

Comments on microbiological testing

_____

_____

_____

Comments on microbiological QA

_____

_____

_____

5. Roles in quality:
   a. Clear and concise roles.
   b. The quality department assumes the role of trainer, not doer of the tests that are close to operations.
   c. Operations conducts the tests that are easy and repeatable, such as pH.
   d. Operation teams have established their accountability to control products within specification, and react before finished goods are impacted by poor quality.
6. New products:
   a. Are planned with some consideration to risk in design and process.
   b. Modifications are made to quality systems to react to changing controls.
   c. The new product development process is owned by someone with responsibility in the quality department.
   d. New products are introduced with consideration of quality risks and production constraints.
7. Operational control points or key process indicators (KPIs) and quality control points (QCPs):
   a. Routinely posted and reviewed as part of operational standing meetings.
   b. The quality or operations departments are responsible for reviewing and communicating status of data.
   c. We control our process inputs such as pressures, temperature, steam rates, and gassing by monitoring the gauges and plotting these on a control chart.
   d. We control our product quality by measuring specifications such as alcohol, bitterness, color, and taste. We plot these on a control chart.
8. SOPs:
   a. Are written for all operations.
   b. Are updated and frequently improved.
   c. Are part of operations and training.
9. The quality and operations departments share responsibility to conduct training for quality testing procedures.
10. Food safety risk assessments are part of process implementation and standing processes are routinely reviewed for risk.

MAINTENANCE AND GMPs/SANITATION
QA and Continuous Improvements
1. Calibration of test equipment:
   a. Check standards and employee accuracy.
   b. Written procedures and reaction levels.
   c. We have structured problem-solving processes:
   d. Problem-solving training module as part of general training.
   e. We have clear roles in problem solving.
   f. We have clear prioritization of what problems to solve.
   g. Management is involved in our day-to-day problems.
   h. We use data and statistical applications to help solve problems.
   i. We have a standard repeatable process from shop floor to higher-level management in order to solve problems.
   j. The quality team helps train the operations on statistics.
2. Preventive maintenance:
   a. We have a master sanitation plan. All areas of the brewery are attended to in terms of cleanliness.
   b. We perform routine inspections and proactive maintenance.
   c. Our team has basic training for maintaining our equipment.
   d. Preventive maintenance schedules are developed and audited.
3. Stores control:
   a. We have non-cluttered work areas.
   b. Work areas are cleaned, organized, and audited.
4. Good manufacturing processes (GMPs):
   a. GMP policy is fully in place.
   b. Training has been conducted.
   c. Brewery is following GMP policy.
5. We have sanitation SOPs developed:
   a. Our GMPs and sanitation programs are verified on a routine basis.

APPENDIX J: GENERAL AUDIT REPORT

    b. Our work areas are clean and we have continuous training of new employees to keep it that way.
6. CIPs:
    a. Are verified by our chemical supplier.
    b. Frequency and efficiency audited with microbiology tests or ATP.
    c. Documented and chemical residues are tested.

GLOSSARY

Acceptance quality level (AQL) – The lowest level of quality that is acceptable based on statistics. Users select a risk level of rejecting a good lot and/or the risk level of accepting a bad lot. Based on this accepted quality level, a plan can be constructed.

Acceptance sampling plan – A plan used to determine how much to sample based on the AQL.

Accuracy – How close a value is to its true value.

Asset care – The maintenance of equipment.

Assignable causes – The special variation caused by unpredictable phenomena in a process, such as energy surge or valve failure.

Centralized quality – Governance or rule setting for quality policy and procedures is centralized to an individual or a small team.

Corrective action – Corrective action is the adjustment to a process after a specification breach.

De-centralized quality – Governance or rule setting for quality policy and procedures is de-centralized and many individuals set separate policy and procedures at multiple sites. This may be necessary if a brewery conglomerate has many different types of brewing operations (e.g., large, small, lager, ale, wild beer, barrel aging, etc.) that a one-size-fits-all set of quality check procedures does not apply to.

ISO-9000 – Internationally recognized set of guidelines and criteria to formalize a quality system.

Lean management – A system of management that became popular in the 1990s when Toyota strongly influenced quality management. Lean management is a system of eliminating waste in an operation.

Level of confidence – A level that expresses in percentage how certain you can be in a statistical test. 95% or 99% is usually selected.

Natural cause variation – The natural variation expected in a process due to predictable phenomena always present in the process.

Precision – How close a value is to itself after repeated testing.

Process capability – Process capability is a calculated statistical value that determines if a process is capable of meeting a specification.

Process input measure – The measurement of all process inputs. In kettle boil, a process input is a steam flow rate.

Process logic controller (PLC) – An automated controller that monitors variable data (such as temperature) and controls (opens or closes) valves or other mechanical devices to maintain control within a certain set of parameters.

Process output check or measure – The measurement or check that is conducted at the end of a process step. In a kettle boil the outputs are the extract and evaporation percentage.

Quality assurance – Checking the quality control measurement system by calibrations, checking standards, or other means.

Quality at the source (QAS) – A quality check that is conducted at the source or process site; for example, pH measurement of wort on the brew deck.

Quality control – Measurement conducted to check product or process while the process is occurring. Conducted as either a process input or a process output.

Quality governance – The setting of policy, strategy, specifications, and goals for the brewery as they pertain to quality of the products.

Quality management – Executing the quality policy and procedures via the measurement and control plan (or the quality system), resulting in process and product control, and great tasting beer.

Quality manual – A quality manual is a written set of policies, procedures, and formal specifications for the brewery to follow.

Quality policy – Policy set by leadership on their expectations of how employees and managers will behave to manage their quality outputs.

Quality system – A quality system is the complete set of governance, management, programs, policies, and procedures that allows for consistency and quality in a company's products.

Reactive maintenance – Reacting after equipment fails or totally breaks down and stops working.

Repeatability – The ability for one analyst to repeat a measurement outcome using one tool on one sample (non-destructible sample). For example, reading pH multiple times on one sample.

Reproducibility – The ability for multiple analysts to repeat a measurement outcome with a measurement tool.

Risks – The inherent errors that can occur as process or product changes are added.

Root cause – The true cause of a quality issue. Once a root cause is fixed, the brewery does not have a repeated failure.

Six Sigma – A formal program that establishes training, governance, and statistical relevance to problem solving.

Specification – A single, specific measurement, usually of a process output, that requires corrective action if breached.

Standard – A set of criteria that formalizes acceptability of a product or raw material.

Statistical process control (SPC) – A statistical calculation that allows a brewery operator or quality technician to determine if process is "in control" or running within the variation expected.

Total quality management (TQM) – A quality system designed to bring all employees a tool to manage their quality outputs. TQM became a popular way to manage quality in the 1980s.

RESOURCES

These are just a few of the many books and web resources that can be referenced for technical brewing or quality management needs.

HACCP and GMP Example Plans

MBAA – http://www.mbaa.com/brewresources/foodsafety/haccp/Pages/default.aspx

Beer Canada – http://www.beercanada.com/beer-canadahaccp-food-safety-program

HACCP, Food Safety, or Safe Quality Foods

American Society for Baking – http://www.aibonline.org/

International HACCP Alliance – http://www.haccpalliance.org/

National Sanitation Foundation (NSF) – http://www.nsf.org/training-education/training-food-safety/

Safe Quality Foods Institute – http://www.sqfi.com/training-centers/

Silliker/Merieux Laboratories – http://www.merieuxnutrisciences.com/

Microscopy Resources

McCrones Research Institute – http://www.mcri.org/v/10/Courses

Positive Controls Supply

Microbiologics – http://www.microbiologics.com/site/index.html

ATCC Control Authority

http://www.atcc.org/Standards/Standards_Programs/ATCC_Standards_Resource/ATCC_Licensed_Derivative.aspx

Quality Control Statistics and Control Charting

Local ASQ – Local ASQ councils provide excellent and inexpensive basic training for statistics. Check individual websites for details.

Minitab – An excellent tool for quality control statistics. Training program is also relevant. http://www.minitab.com/en-us/training/courses/

National ASQ – http://asq.org/training/introduction-to-statistical-concepts_ISCASQ.html

International Commission for Microbiological Specifications in Food (ICMSF) – Book 2 is available as a free download at the ICMSF web site. http://www.icmsf.org/pdf/icmsf2.pdf

Quality Control Statistics Computer Packages

Minitab – The gold standard. Excellent tool for Six Sigma plant. http://www.minitab.com/en-us/

QI Macros – Runs off Excel data. A good tool to get started with. http://www.qimacros.com/spc-software-for-excel/statistical-quality-control-software//

Quality Manager Certification

ASQ http://asq.org/cert/manager-of-quality

American Society for Quality (ASQ)
http://asq.org/learninginstitute/index.html or access the local ASQ chapter for references in your area.

Quality Management and Systems

ASQ Quality Progress Magazine – A monthly publication, free to members. http://asq.org/qualityprogress/index.html

Taque, Nancy R. 2005. *The Quality Toolbox.* Milwaukee WI: American Society for Quality, Quality Press.

Westcott, Russell T. Ed. 2005. *Certified Manager of Quality and Organizational Excellence, 3rd ed.,* Milwaukee, WI: American Society of Quality, Quality Press.

Sensory Analysis Tool Kits

Aroxa – https://www.aroxa.com/beer-guide

Cicerone off-flavor kits and training – http://cicerone.org/off-flavor

Flavoractiv – http://www.flavoractiv.com/drinks/beers/beer-flavour-standards/how-use-gmp-flavour-standards/

Siebel Institute – https://www.siebelinstitute.com

Six Sigma

ASQ – http://asq.org/cert/six-sigma-black-belt

Villanova University – http://www.villanovau.com/lp/six-sigma/career/cert-mst_leanblackbelt_t01_1101/?mcguid=-ca75c245-b0f0-4272-aa64-71feb3196f28&mcid=21459

Check with your local technical college or search for online programs.

Technical Brewing Resources

American Society of Brewing Chemists. 2007. *Methods of Analysis of the American Society of Brewing Chemists.* http://www.asbcnet.org/MOA/

Bamforth, Charles. 2009. *Beer: A Quality Perspective.* Minneapolis, MN: Master Brewers Association.

Boultan, Chris and David Quain. 2006. *Brewing Yeast and Fermentation.* Oxford UK: Blackwell Sciences Publishing.

Edwards, C.G. 2005. *Illustrated Guide to Microbes and Sediments in Wine, Beer, and Juice.* WineBugs LLC, Pullman, WA.

Kunze, Wolfgang. 1999 *Technology of Brewing and Malting.* Berlin, Germany: Versuchs und Lehrnstalt für Brauerei.

McCabe, John. 1999. *The Practical Brewer.* Minneapolis, MN: Master Brewers Association.

Ockert, Karl and Ray Klimovitz. Eds. 2014. *Beer Packaging, 2nd ed.,* Minneapolis, MN: Master Brewers Association.

Priest, Fergus G. and Iain Campbell. Eds. 2003. *Brewing Microbiology, 3rd ed.,* New York, NY: Springer Science

Speers, Alex. Ed. 2012. *Yeast Flocculation, Vitality, and Viability. Proceedings of the 2nd International Brewers Symposium.* Minneapolis, MN: Master Brewers Association.

BIBLIOGRAPHY

American Society of Brewing Chemists. 2001. *Methods of Analysis, 8th ed., Microbiology Yeast – 3*, "A Dead Yeast Cell Stain (International Method)," St. Paul, MN.

American Society of Brewing Chemists. 2001. *Methods of Analysis, 8th ed., Microbiology Yeast – 6*, "Yeast Viability by Slide Culture," St. Paul, MN.

American Society of Quality. 2013. *Global State of Quality*.

Anderson, R. G. 1992. "The Pattern of Brewing Research: A Personal View of the History of Brewing Chemistry in the British Isles." *Journal of the Institute of Brewing* Vol. 98, no. 2.

Bank, John. 1992. *The Essence of Total Quality Management*. Englewood Cliffs, NJ: Prentice Hall.

Bamforth, Charles W. 1985. "Biochemical Approaches to Beer Quality." *Journal of the American Society of Brewing Chemists*. St. Paul, MN: American Society of Brewing Chemists. May-June, Vol. 91.

Bamforth, Charles W. 2002. *Standards of Brewing: A Practical Approach to Consistency and Excellence*. Boulder, CO: Brewers Publications.

Baron, Stanley. 1962. *Brewed in America: The History of Beer and Ale in the United States*. Boston: Little Brown.

Brenner, M. W. 1996. "Some thoughts on quality control." *MBAA Technical Quarterly* St. Paul, MN: Master Brewers Association of the Americas. Vol. 33, no. 3. (Republished from 1953).

Brewers Association. 2014. Brewers Association FDA and FSMA Supplemental Comment. https://www.brewersassociation.org/wp-content/uploads/2014/12/FDA-FSMA-Supplemental-Comment.pdf. Accessed 05/20/15.

Carter, Steven and Jeremy Kourdi. 2003. *The Road to Audacity: Being Adventurous in Life and Work*. New York: Palgrave McMillan.

Center for Food Safety and Applied Nutrition. Food and Drug Administration. 2012. "Guidance for Industry: What You Need to Know About Registration of Food Facilities Small Entity Compliance Guide." Office of Compliance, HFS-607. College Park, MD 20740

Chilver, M. J., J. Harrison, and T. J. B. Wevv. 1978. "Use of Immunofluorescence and Viability Stains in Quality Control." *Journal of the American*

Society of Brewing Chemists. St. Paul, MN: American Society of Brewing Chemists, 36:13–18.

Christian, J. H. B., and T. A. Roberts. 1986. *Microorganisms in Foods 2: Sampling for Microbiological Analysis: Principles and Specific Applications.* Toronto: University of Toronto Press.

Corran, H.S. 1975. *A History of Brewing.* North Pomfent, VT: David and Charles, Inc.

Del Castillo, Enrique. 2002. *Statistical process adjustment for quality control.* New York: Wiley.

Dick, Steven J., and Roger D. Launius. 2007. *Societal impact of spaceflight.* Washington, DC: National Aeronautics and Space Administration, Office of External Relations, History Division. http://purl.access.gpo.gov/GPO/LPS115997.

Earle, Mary D., Richard L. Earle, and Allan M. Anderson. 2001. *Food Product Development.* Boca Raton, FL: CRC Press.

Goff, R. E. 1995. "TQM at Anheuser-Busch... Tomorrow's Quality Management." *Brewers Digest* 70, 4.

Gurda, John. 2006. *Miller Time. A History of Miller Brewing Co. 1855–2005.* Milwaukee, WI: Miller Brewing Co. 134, 156.

Hart, Marilyn K. and Robert F. Hart. 1989. *Quantitative Methods for Quality and Productivity Improvement.* Milwaukee, Wisconsin: American Society for Quality, Quality Press.

Hull, TC. 1990. "Reducing costs in the brewing industry through total quality management." *MBAA Technical Quarterly* St. Paul, MN: Master Brewers Association of the Americas. Vol. 27, no. 2.

Hutchinson, Donald J. 2012. "The Measurement of Carbon Dioxide in Packaged Beer: A critical review." Abstract. World Beer Congress 2012, Technical Session 09: Analytical II Session.

Hutter, K. J. 1999. "Control of Cell Cycle, Glycogen Content and Viability of Brewing Yeasts During Fermentation by Flow Cytometric Analysis." Monograph—European Brewery Convention. 28: 194–197.

Hutter, K. J. and C. J. Lange. 2001. "Yeast Management and Process Control by Flow Cytometric Analysis." Proceedings of the European Brewing Convention.

Keats, J. Bert, and Douglas C. Montgomery. 1996. *Statistical Applications in Process Control.* New York: M. Dekker.

Kara, B. V., W. J. Simpson, and J. R. M. Hammond. 1988. "Prediction of the Fermentation Performance of Brewing Yeast with the Acidification Power Test," *Journal of the Institute of Brewing.* 94: 153–158.

Kohlmann, Frederick J. 2003. "What Is pH and How Is It Measured?" *A Technical Handbook for Industry.* Hach Company, GLI. LITG004.

Kunze, Wolfgang. 1999. *Technology of Brewing and Malting.* Berlin, Germany: Versuchs und Lehrnstalt für Brauerei.

Loch, C., and Stylianos Kavadias. 2007. *Handbook of new product development management.* Amsterdam: Butterworth-Heinemann.

McCabe, Jack. 1995. "Brewing! Have Things Changed So Much? Brewing "esoteric"? Methods "old fashioned"? "Master brewer" displaced by "chemist"?" *MBAA Technical Quarterly* St. Paul, MN: Master Brewers Association of the Americas. Vol. 32, no. 4, 241–243. (Republished from 1948).

McCabe, Jack. 1995. "American Brewing Operations Past and Future." Paper presented by Gustave Geob. *MBAA Technical Quarterly* St. Paul, MN: Master Brewers Association of the Americas. Vol. 32, no. 3, 115–118. (Republished from 1948).

Mao Sugihara, Takeo Imai, Masahito Muro, Youko Yasuda, Yutaka Ogawa, Motoo Ohkochi. "Application of Flowcytometer to Analysis of Brewing Yeast Physiology." Research Laboratories for Brewing, Kirin Brewery Co., Ltd.

Bouix, Marielle and Jean-Yves Leveau. 2001. "Rapid Assessment of Yeast Viability and Yeast Vitality During Alcoholic Fermentation." *Journal of the Institute of Brewing*. Vol. 107, no. 4, 217–225.

McCaig, R. 1990. "Evaluation of the Fluorescent Dye 1-anilino-8-naphthalenesulphonic Acid for Yeast Viability Determination." *Journal of the American Society of Brewing Chemists*. St. Paul, MN: American Society of Brewing Chemists. 48: 22–25.

Menz, G., P. Aldred, F. Vriesekoop. 2011. "Growth and Survival of Foodborne Pathogens in Beer." *Journal of Food Protection*. Oct. 74 (10):1670-5. doi: 10.4315/0362-028X.JFP-10-546.

Mochaba, F., E. S. C, O'Connor-Cox, and B. C. Axcell. 1998. "Practical Procedures to Measure Yeast Viability and Vitality Prior to Pitching," *Journal of the American Society of Brewing Chemists*. St. Paul, MN: American Society of Brewing Chemists. 56:1–6.

Mochaba, F., E. S. C, O'Connor-Cox, and B. C. Axcell. 1997. "A Novel and Practical Yeast Vitality Method Based on Magnesium Ion Release," *Journal of the Institute of Brewing*. 103: 99–102.

Munro, Roderick A., Govindarajan Ramu, and Daniel J. Zrymiak. 2015. *The Certified Six Sigma Green Belt Handbook*. Milwaukee, WI: ASQ Quality Press.

O'Connor-Cox, E. S. C., F. M. Mochaba, E. J. Lodolo, M. Majara, and B. C. Axcell. 1997. "Methylene Blue Staining: Use At Your Own Risk". *MBAA Technical Quarterly*. St. Paul, MN: Master Brewers Association of the Americas. 34 (1): 306–312.

Paar, Anton. 2015. SS-60 Volume Meter (SERIES 1000) http://www.zahmnagel.com/Products/SERIES1000/tabid/60/.

Paar, Anton. 2015. Alcolyzer Beer Analyzing System (:: Anton-Paar.com) http://www.anton-paar.com/us-en/products/details/alcolyzer-beer-analyzing-system/beverage-analysis/ Accessed 05/20/2015.

Peddie, F. L., W. J. Simpson, B. V. Kara, Sarah C. Robertson, and J. R. M. Hammond. 1991. "Measurement of Endogenous Oxygen Uptake Rates of Brewers' Yeasts." *Journal of the Institute of Brewing*. 97 (1): 21–25.

Priest, Fergus G., and I. Campbell. 2003. *Brewing Microbiology*. Boston, MA: Springer US. http://dx.doi.org/10.1007/978-1-4419-9250-5.

Raju, Tonse N. K., MD. 2005. "William Sealy Gosset and William A. Silverman: Two 'Students' of Science." *Pediatrics* 116 (3).

Roberts, Elisabeth and Chris Klein. 2007. "Improved Method for CO2 Measurements." *Brewing and Beverage Industry International* (reprint) Vol. 5. Verlag W. Sachon: Schloss Mindelburg.

Ryan, Thomas P. 2000. *Statistical Methods for Quality Improvement*. New York: John Wiley and Sons.

TAD. "TTB | TAD | Common Compliance and Tax Issues During Brewery Audits." Overviews & Factsheets. Accessed May 20, 2015. http://www.ttb.gov/beer/beer-tutorial.shtml.

Tennant, Geoff. 2001. *Six Sigma: SPC and TQM in manufacturing and services*. Aldershot, England: Gower.

Tenny, Robert I. and Philip E. Dakin. 1984. "History of the American Society of Brewing Chemists." *Journal of the American Society of Brewing Chemists.* St. Paul, MN: American Society of Brewing Chemists. 42:0098, 1984.

Trembley, Victor and Carol H. Trembley. 2005. *The U.S. Brewing Industry: Data and Economic Analysis.* Boston: Massachusetts Institute of Technology.

Trevors, J. T., R. L. Merrick, I. Russell, and G. G. Stewart. February 1983. "A Comparison of Methods for Assessing Yeast Viability." *Biotechnology Letters*, Vol. 5, no. 2, 131–134.

Skilnik, Bob. 2006. *A History of Brewing in Chicago.* Ft Lee, NJ: Barricade Books.

Smart, K. A., K. M. Chambers, I. Lambert, & C. Jenkins. 1999. "Use of Methylene Violet Staining Procedures to Determine Yeast Viability and Vitality." *Journal of the American Society of Brewing Chemists.* St. Paul, MN: American Society of Brewing Chemists. 57: 18–23.

Swanson, Roger C. 1995. *The Quality Improvement Handbook: Team Guide to Tools and Techniques.* Delray Beach, FL: St. Lucie Press.

Van Zandycke, S. M., O. Simal, S. Gualdoni, and A. Smart. 2003. "Determination of Yeast Viability Using Fluorophores." *Journal of the American Society of Brewing Chemists.* St. Paul, MN: American Society of Brewing Chemists. 61:15–22.

Vrellas, Charisis G., George Tsiotras. 2015. "Quality Management in the Global Brewing Industry." *International Journal of Quality & Reliability Management.* Vol. 32: 1 42–52.

Westcott, Russell T., Ed. 2005. *Certified Manager of Quality and Organizational Excellence, 3rd ed.,* Milwaukee, WI: American Society of Quality, Quality Press.

INDEX

Entries in **boldface** refer to photos and illustrations.

Accelerated Batch Fermentation (ABF), 13
Acceptance quality level (AQL), 43; defined, 159
Acceptance sampling plans, 43-44; defined, 159
Accountability, 15, 18, 25, 30, 51, 59
Accuracy, 5, 30, 70, 84, 86
Adenosine triphosphate (ATP), 57, 70
Adjustments, 39, 41, 62, 64, 69
Alcohol, xii, 38, 64, 86; fermentation results of, 40; measuring, 83-84, 85; meters, 70; monitoring, 100-101; oxidation, 85; percent, 6; removing, 78
Alcolyzers, 119
American Society of Brewing Chemists (ASBC), 12, 34, 44, 72, 74, 75, 83, 86, 87
American Society of Quality (ASQ), 17, 25, 162; foundation training and, 30; Juran and, 4; reference by, 32; Six Sigma and, 31
Amperometric, 94, 95
Analysis, 16, 31; sensory, 86-90
Analytical and sensory sampling program, **124-126**
AQL. *See* Acceptance quality level
ASBC. *See* American Society of Brewing Chemists
ASBC Method Microbiological Control 6: 57
ASBC Methods of Analysis, 82, 83, 88, 101
ASBC Microbiological Control-5, 81
ASQ. *See* American Society of Quality
Assessment: department, 104, 106; food safety risk, 156; product, 104, 106, **108**, 109; quality, 103. *See also* Risk assessment

Assets, 48; caring for, 31, 50, 53, 159; maintenance for, 51-54
Assignable causes, 39; defined, 159
ATCC control authority, resources for, 161
Attenuated total reflectance (ATR), 92-93, 94
Audits, 9, 55, 98, 153-157; communication and, 105; compliance, 104; conducting, 104, 105; conformance, 104; formal, 99; quality system, 103, 105-106, 109, 153; structured, 19, 103; system, 104; third-party, 105; traceability, 99, 101; types of, 104-105
Autoclave operation, 78
Automation controls, 32

BA. *See* Brewers Association
Bacteria, 29, 54, 62, 79, 144; lactic acid, 81; staining, 74; types of, 63; viewing, 71, 72, 73
Balling tables, 83, 84, 85
Beer and pressure/temperature chart, carbon dioxide solubility in, **93**
Beer Canada, 36
Beer gas, determining, 91
Beer-Lambert Law, 85
Bioterrorism Act (2002), 98
Bitterness, xii, 7, 37, 38, 51, 64
Bitterness unit (BU), 37, 38
Bloomberg Businessweek, 11
Body of knowledge (BOK), 25

Bottle conditioning, 33
Bottle volume, 6
Bottles: glass, 146; Nalgene, 87
Brand quality, xvii, 7
Brenner, Mortimer, 3, 11
Brettanomyces, 60
Brewers Association (BA), 5, 12, 101
Brewer's Laboratory Handbook (Brewing Science Institute), 82
Brewery managers, 67, 105
Brewhouse vessels, sanitation and, 113
Brewing, 5, 13, 17, 32, 47, 51, 154; science/art of, xii, 7, 11; techniques/reviving, xviii
Brewing Science Institute, 82
Brewmasters: risk failure and, 35; role of, 10, **10**, 11, 14, 16, 17, 20, 113
Brightfield methylene blue/violet stain test, **75**
Brix, 84
BU. *See* Bitterness unit

Calendar of Innovations Strategy, example of, **61**
Calibration, 30, 32, 44, 45, 62, 92, 94, 100, 101; monitoring, 22; routine, 91
Carbon dioxide, 43, 70; checks, 91, 92; dissolved, 91, 94; levels, 23, 29, 91; measuring, 92, **92**; risks/benefits of, **94**; solubility, **93**; at Standard Temperature and Pressure, 92
Carrot Principle, The (Gostick), 51
Caustics, 87, 95, 114
Cell density, 71, 75, 76
Cell staining, 72-73
Centralized quality, defined, 159
Certificate of Analysis (COA), example of, **63**
Certification for Quality Technician and Quality Manager (ASQ), 32, 50
CFU. *See* Colony forming units
Change, 2, 9, 21; incremental, 60, 61; process, 62, 104; radical, 61
Checks, 90; carbon dioxide, 91, 92; frequency of, 41, 44; go/no-go, 43, 88; gravity, 45; iodine, 8, 42; maintenance, 141; master list of, 30; microbiology, 42, 44; packaging, 43; pH, 23; QA, 74, 91; QC, 44; transition, 30. *See also* Quality checks
Chemicals, 56, 57, 89, 100, 140; cleaning, 87; pesticide, 145; residual, 34, 86

Chemistry, 82, 86; tests/go/no-go sensory/analytical, 43
Chemists, 16, 29, 75; role of, 11-12, 14
Chief operating officer (COO), quality and, 21
Chill-haze, 7
CI. *See* Continuous improvement
CIP. *See* Clean-in-place
Clarity, xii, 6, 13
Clean-in-place (CIP), 18, 53, 78, 94, 104, 157; procedures for, 57; responsibility for, 113
Cleaning, 53, 55, 57, 67, 73, 87, 100, 113, 115, 139, 144-145, 146; acid, 114
Colony forming units (CFU), 42
Color, xii, 1, 7, 38, 64, 85, 86
Communication, 2, 8, 45, 56, 58, 62, 64, 66, 99, 155; audits and, 105; continuous, 25; maintenance and, 53; metrics, 14; organizing, 25; problem-solving, 31, 67; quality and, xii, 16; timely, 154
Construction contractors, 146-147
Consultants, audits and, 105
Consumer complaints, 18, 35; logging, 57-58, 116, **118**
Consumers: criteria, 14; education, 7; expectations of, 1, 28; respect for, 19
Contamination, 53, 145, 146, 147
Continuous improvement (CI), 14, 19, 24-26, 67, 156
Contractors, 55, 147
Control, 4, 31, 45; culture of, 15; empowerment of, 14; level of, 42; sanitary, 55; system, 10, 42
Control charts, 4, **38**, 40, **40**; data on, 43; resources for, 161-162
Control limits, 33, 34, 38, 45, 64; setting, 37, 39-41
Corrective actions, 22, 23-26, 27, 45, 112; defined, 159; records for, 57; short-term, 105
Counting chambers, 76, 77
Counting technique, poor, 77
Cpk. *See* Process capability
Craft brewers: growth of, xi, 6; opportunity and, xv; quality and, xii, xvii
Critical Control Point Justification, **132**
Critical Control Point Monitoring Summary, **133**
Critical Control Points, **114**, 131-133, 156
Crosby, Philip B., 3, 5, 12; fitness for use and, 6; TQM and, 4, 7
Culture, 15, 16, 50, 55-56, 67, 99; of quality, xii, xv, xvi, 2, 14, 24, 28, 68

Data, 32, 39, 97; access to, 43; accuracy of, 30; analyzing, 9, 31, 43, 66, 67, 88; limiting, 42; maintenance, 52; monitoring, 121-122, 141; pre-control, 64; QC, 44, 52; records/templates for, 57; sharing, 25; statistics and, 31

De-centralized quality, defined, 159

Decision making, 16, 18, 19; data-driven, 41; delegating, 17; management, 69; quality-related, 20

Define, Measure, Analyze, Improve, Control (DMAIC), 31, 32, 66

Degrees Plato, 23, 29, 83, 84, 146

Deming, W. Edwards, 3, 4, 5, 8, 9, 39

Density meters, 70

Diacetyl, 70, 87

Dilution, 76, 78, 82, 87

Disease control, 55, 144

Dissolved oxygen (DO), 3, 39, 43, 53, 90, 95, **95**, 146; control chart of, 38; liquid, 91; measuring, 94

Distributors, 47, 51

DMAIC. *See* Define, Measure, Analyze, Improve, Control

DO. *See* Dissolved oxygen

Documentation, 8, 62, 100, 104, 111; formal, 99; sanitation/CIP policy and, 113

Drains, cleaning, 144, 145

Dry hopping, 59, 60

Duo-trio test, **88**; described, 88

Efficiency: cost, 25; impact on, 52; quality and, 19, 25, 51

Employees. *See* Staff

Engineering, 9, 51, 53

Equipment, 45, 62; basic setup for/brewery size, 70; calibrating, 30; CIP, 113; cleaning/sanitizing, 53, 100, 144-145, 146; critical process, 53; designing, 54; installing, 52, 54; maintaining, 22, 52, 55, 61; manual offline vs. in-line CO_2 testing, **92**; quality control check, 44; record keeping for, 100; storing, 144

Escalation process, 20, 21

European Brewers Convention (Analytica-EBC), 83

Evaluation, quality, 45

Excise tax laws, 97

Expectations: consumer, 1, 28; setting, 15-21, 24, 31

Extract, measuring, 83-84

Failure modes, 36, 90, 91, 94

Failure Modes and Effect Analysis (FMEA), 37, 52; example of, 135-136, **136**; risk assessment and, 36, 61, 62

FDA. *See* Food and Drug Administration

Feedback, 64, 154

Fermentation, xiii, 5, 8, 23, 29, 79, 122, 135; accelerated, 13; complete, 35; issues, 51; process, 129; rate, 40

Fill levels, monitoring, 91, 100-101

Filtration, 78, 91, 95

Finished Beer Specification, 122; example of, **130**

Finished goods sampling plan, **127**

First-in and first-out (FIFO), 155

Fitness for Use, quality as, 4, 6-7

Flavor: adding, 59; bandaid, 104; fermentation results of, 40; hop, 89; inconsistent, 13; off, 6, 86, 87, 88, 89; on, 89; optimizing, 61; profile, 6, 62; sensitivity, 11; sour, 38; stability, 33

Fluorescent dye test, **75**

FMEA. *See* Failure Modes and Effect Analysis

Foam, 1, 38; retention/inconsistent, 13

Food, 119; adulterated, 54; quality and, 7

Food and Drug Administration (FDA), 36, 54, 97, 99, 101; authority of, 98; GMP policy and, 143

Food plants, registration of, 98

Food poisoning, 144

Food safety, 34, 35, 36, 45, 52, 53, 60; controlling, 97, 137; microbiological, 1; resources for, 161; risk assessment and, 45, 97-101, 156; rules for, 97

Food Safety Hazard Analysis, **132**

Food Safety Modernization Act (2011), 36, 98, 101

Free amino nitrogen (FAN), 85

Free from defect, quality as, 5-6

Gage Reliability and Repeatability (Gage R&R), 22

Gas chromatograph, 23, 70, 83

Gas diffusion membrane, 92-93

Gas testing, 45

General audit report, 153-157

General Motors, 8

Glass, 34, 90, 91, 119; breakage, 115; breakage log, **115**; cleaning, 146; control, 55; evacuation, 115

Glass policy, described, 115

Global State of Quality (ASQ), 17

Glycol, 39
GMP. *See* Good Manufacturing Plan
GMPs. *See* Good Manufacturing Processes
Goals, 15, 155; lab, 69; problem-solving, 66; quality, 51, 57-58, 66, 68; setting, 4, 17-19, 51, 57-58
Good Manufacturing Plan (GMP), 55; example/resources for, 161
Good Manufacturing Processes (GMPs), xii, xviii, 27, 54-57, 100, 101, 113, 114, 115, 156-157; budget for, 56; culture of, 55-56; employee training and, 99; example of, 143-147; facility grounds, 144; foundational requirement of, 98; implementing, 55-56, 99; inspections, 98, 99, 155; sanitary operations, 144-146
Gostick, Adrian, 51
Governance, 67, 112-113; centralization/decentralization of, 16, 17; defined, 160; expectations of, 15-21; management, 15-16, 26, 27; mega-brewery, **19**; mid-size brewery, **19**; quality management and, 15, 16; roles/responsibilities of, 18, 20, 112; small craft brewery, **19**
Gram staining, 72, 73, 81; mistakes with, 74; technique for, **73**
Gravity, 29, 83, 84; checks, 45; target, 64
Growth, xi, xv, 6, 14, 33, 34, 81; media, 79; quality and, xviii
Guinness & Company, 9

HACCP. *See* Hazard Analysis and Critical Control Points
Hand washing, 100, 144, 145-146
Hazard Analysis and Critical Control Points (HACCP), xii, xvii, 27, 35, 36, 37, 50, 54, 101; described, 4; Process Map, 137-138, **138**; resources for, 161; risk assessment and, 34, 131-133
Haze, 7, 33
Head brewers, roles/responsibilities of, **10**
Hemocytometers, 70, 75, 76, 82; counting grid, **76**; math example, 77
Henry's law, 91, 92, 94
High performance liquid chromatography (HPLC), 83
HLP test. *See* Hsu's lactobacillus/pediococcus test
Hoods, 70, 80

Hop alpha acids, 38
Hops, 6, 83, 89
Hsu's lactobacillus/pediococcus (HLP) test, described, 81
Hull, T.C.: TQM and, 13
Human resources (HR), 2, 17, 47-51, 52, 57, 155; accountability and, 59; coordination by, 56; goal setting and, 51
Human resources (HR) manager, quality and, 50
Hydrometers, 41, 84
Hygiene, 100, 144

Illnesses: food borne, 36; signs of, 144
Illumination: brightfield, 71; Koehler, 71, 72; optimal, 71
Implementation, 21-26, 29, 55-56, 59-62, 64, 66, 99
Improvement, xiii, 4, 15, 16, 31, 41, 64, 66, 109; continuous, 14, 19, 24-26, 67, 156; core-capability, 60; incremental, 60; process of, 5; radical, 60, 61; standardizing, 25; team, xviii; TQM and, 13
Incubation, 78, 79
Inductively Coupled Plasma (ICP), 83
Innovation, xi, 14, 19, 62; incremental, 60; quality and, xviii; radical, 60, 61; strategy for, 60
Insecticides, 145
Inspections, 9, 55, 98, 115
Instrumentation, 85, 94
International bitterness units (IBUs), 85
International Commission for Microbiological Specifications in Food (ICMSF), 42
International Organization for Standardization (ISO), 25, 104
International Telephone and Telegraph, Crosby and, 4
Inventory, 57, 99, 155
Iodine conversion, 41, 42
ISO-9000, 50; defined, 4, 159
ISO. *See* International Organization for Standardization

JE Siebel and Sons, 81
Juran, Joseph M., 3, 4, 5, 8, 9, 12

"Keeping Qualities of Beer, The" (Tenny), 12
Kegs, cleaning, 67

Key process indicators (KPI), 156
Knowledge, 2, 6, 25, 49, 52; statistics, 31; technical, 11, 29
Labels, 6, 90, 97, 122
Labor issues, 12, 14
Laboratories, 76, 82, 101, 119; described, 70; microbiological media in, **80**; space requirements for, 70
Laboratory Handbook (Brewing Science Institute), 82
Laboratory managers, 69
Laboratory manuals, 82
Laboratory technicians, 28, 49, 62; learning matrix for, **49**
Lactobacillus, 60, 72, 81, 82
Leadership: development program, 50; localized, 17
Lean Management, 50; defined, 4, 159
Learning, 19, 48, 49, 100; matrix, **49**
Learning and development (L&D), managing, 48
Level of confidence, defined, 159
Limits: control, 33, 34, 37, 38, 39-41, 45, 64; reaction, 37, 41; rejection, 45; setting, 37, 39-41, 45
Lock out tag out (LOTO), 140
Lubrication, food-grade, 53, 100

Maintenance, 31, 44, 55, 94, 98, 141, 146-147, 156-157; accountability and, 59; asset, 51-54; communication and, 53; general, 144; monthly, 95; preventive, 156; proactive, 52; programs, 47, 51-52, 100; quality and, 21, 52-53; quality inspections for, 141-142, **142**; reactive, 160; records, 57; scheduled, 45
Maintenance managers, 53; sanitation and, 56-57
Malt, 6, 106
Management, 29, 48, 53, 56, 69, 103, 105, 153; focus on, 3; gaps in, 2; governance and, 15-16, 26, 27; influence and, 32; layers of, 48; media, 77-78; middle, 30-32; orientation process, 23; philosophies, 3;
policy changes and, 21; quality tasks in, 20; regulatory compliance, 22; skills for, 30-32, 47, 48. *See also* Quality management
Managerial Breakthrough (Juran), 4
Manual offline vs. in-line CO_2 testing equipment, **92**
Manufacturing, 28, 113, 146; quality, 2, 3
Mash, 8, 9, 22, 36, 42; pH of, 86

Mash tun, worst-case position for, 43
Master Brewers Association of the Americas (MBAA), 3, 11, 13, 36, 143; ASBC and, 12; GMP general template from, 55
Master Sanitation Schedule (MSS), 56-57, 113-114
Material Safety Data Sheets (MSDS), 62, 119, 140
MBAA. *See* Master Brewers Association of the Americas
Measurement, 12, 14, 21, 31, 41, 83-84, 92, 94; alcohol, 85; analysis of, 22; control, 45; electrical, 86; fill-volume, 101; instruments/controls for, 146; QC, 129; quality and, xii-xiii; science of, 9; static in-line, 43; system of, 121; temperature-sensitive, 84
Media: inoculation of, 81; management, 77-78; microbiological, 77, **80**; overview of, 82; preparing, 78, 81
Megabreweries, 9, 58
Methods of analysis (MOA), 34
Methylene Blue viability stain test, 74
Metrics, 14, 18, 33, 37
Microbiological plating, 77-78, 79
Microbiology, 1, 29, 38, 51, 57, 62, 70, 71, 72-73, 79, 105; brewing, 63 checks, 42, 44; issues, 35, 53, 73, 78; sampling program, **122-123**
"Microbiology Methods in Brewing Analysis" (Campbell), 79
Microflora, 79
Microorganisms, 42, 63, 71, 72, 73, 78, 79, 82
Microscopy, 70, 71-72, 74, 75, 161
Middle management, skills of, 30-32
Miller archives, 12
Minitab, 31
Molds, 79
Monitoring, 33, 37, 40, 45, 47, 53, 54; alcohol, 100-101; quality, 41, 53; visual, 36
Motorola, 8
MSDS. *See* Material Safety Data Sheets
Mycotoxins, 35

NASA, 36
Natural cause variation, 39, 40; defined, 159
Newsweek, 11, 12
NIR spectrometers, using, 85
Non-Staining KOH Test, 72

Obama, Barack, 101
Operations, 20, 37, 54; planning, 155; quality and, 14; strategy, 155
Operations department, 156
Operations managers, 61; role of, 17; training and, 48
Operations team, 15, 16, 47, 76
Organization chart, **112**
Organization design, 17, 20, 32
Outputs: checking, 42; identifying, 36; quality, 45, 67
Oxygen, 45; ingress, 33; pickup, xiii. *See also* Dissolved oxygen; Total package oxygen

Paar, Anton, 85
Packaging, xi, 9, 17, 32, 47, 51, 59, 60, 91, 105, 141, 145, 154, 155; requirements for, 122
Packaging Equipment, sanitation/CIP policy and, 113
Packaging manager, 115
Parameters, quality, 7, 40
Pathogens, 34, 35
PDCA program, 66
Pediococcus, 35, 60, 81
Performance: metrics, 18; quality and, 8, 16
Pest control, 55, 113, 145
Pests, 119, 144
pH, 29, 45, 146; calibration of, 44; control, 24; mash, 86; measuring, 22, 86; specification, 24
pH meters, 70, 75, 86
pH probes, 23, 128
Philosophy, quality, 16, 28, 52
Phosphoric acid, 113
Pillsbury, 36
PIM. *See* Process input measurement
Pipetting, 70, 78, 79-80, 81, 82, 119
Planning, 4, 23, 48, 53, 56-57, 61, 155; continuous, 25; developing, 66; strategic, 9, 24
PLC. *See* Process logic controller
Policy, 3, 15, 51, 57, 59, 111; adherence/culture of, 99; interpretation, 21; setting, 17-19
Polymerase chain reaction (PCR), 70
Polyphenols, 85
POM. *See* Process output measurement
Positive control, 72, 80, 82, 161
Pre-control phase, 40, 64
Precision, 84; defined, 159
Pressure gauges, 92, 94

Prevention, 54, 115
Priorities, xvi, 31, 37, 56, 66-67
Problem solving, 9, 14, 32, 59; formal, 64, 65, 66; goals for, 66; managing, 66-67; principles of, 31; prioritizing, 31, 66-67; programs for, 66; structured program for, 64-68, **65**; techniques for, 39; training and, 31, 65-66
Procedures, 3, 17, 18, 48, 49, 51, 59, 111, 140; change, 104; gaps in, 105; outlining, 56; recall, 101, 155; sanitation, 56, 57, 114; standard, 21, 61
Process, 24, 34; adjustments, 69; assessment, 35, 104, 106, **106-108**; complexity, 33; establishing, 30; improvement program, 40-41; knowledge, 49; mapping, 36; monitoring, 45; standardized, 14, 95; target, 41; understanding, 62
Process capability (Cpk), 39, 40; defined, 159
Process control, 8, 31, 32, 37, 45; additional, 36; automatic, 40; context of, 41; maintaining, 9
Process Control Plan, 121-122
Process Control Standard, 129; example of, **129**
Process improvement program, 40-41
Process input, 36, 41
Process input measurement (PIM): defined, 159; POM and, 8
Process logic controller (PLC), defined, 160
Process map, 37, 137
Process output measurement (POM): defined, 160; PIM and, 8
Process Standard Plan, 121-122
Process variation, 30, 40, 41
Product assessment, 106, 109; components of, **108**; described, 104
Product control plan, **151**
Product design, 45, 59-62, 64
Product pullback, 53, 64, 117
Product quality, 25, 28, 33, 36, 48, 50, 141; control plan for, 149-151, **150**; issues, 154; loss of, 37, 104; managing, 47; roles/responsibilities for, 154
Product risks, understanding, 62
Production, 47, 48, 115; areas, 147; quality and, 2, 105; records, 99; responsible, xv; system, 99; tracking/tracing, 101
Production manager, 51
Products: introducing, 62, 64; new, 35, 156; quality, 49
Pycnometers, 84

QA. *See* Quality assurance
QAP. *See* Quality assurance plan
QAS. *See* Quality at the source
QC. *See* Quality control
QCP. *See* Quality control plan
QM. *See* Quality managers
Quality, 48, 50, 53, 60, 103; basics of, 33; culture of, xii, xv, xvi, 2, 14, 24, 28, 68; defining, 5-7, 13; ensuring, xi-xii; as esoteric, 7; good, 1, 6; improving, xiii, xv-xvi, xvii, 19, 25, 26, 41, 51, 52, 58, 64, 66, 109; integrating, 54; loss of, 61; maintaining, 8, 9, 23, 112; monitoring, xv, 27, 33, 47, 54, 82; perception of, 3; poor, xvii, 1, 6, 9, 52; production and, 2, 105; questioning, 90; responsibilities and, 12-14; roles in, 156; self-sufficient, 21; supporting, 67; well-established, 15
Quality assurance (QA), xv, 9, 25, 26, 70, 72, 82, 90, 94, 104, 155-156; checks, 74, 91; defined, 160; QC and, 22, 24
Quality assurance plan (QAP), 27, 33, 44, 45, 116, 121-130; example of, **128**
Quality at the source (QAS), 23, 28, 29, 30; defined, 160; quality culture and, 24
Quality checks, 23, 29, 36, 41, 121; adjusting, 62; common, 141; conducting, 24, 28, 52; maintaining, 30; records for, 57; responsibility for, 12; static in-line, 43
Quality control (QC), xi, 8, 9, 13, 26, 33, 45, 87, 95, 104, 155-156; defined, 160; maintaining, 25, 51, 112, 121; program, 72; QA and, 22, 24; raw material, 32; samples of, 91; statistics, 12, 32, 161-162; tools, 86; using, 23
Quality Control Handbook (Juran), 4
Quality control plan (QCP), 33, 44, 47, 70, 115, 116, 121-130, **122**; developing, 62-63; standard, 64
Quality Control Plan with Specifications, Metrics, and Measures, 27
Quality control points (QCPs), 156
Quality department, 17, 22, 23, 28, 60, 156; assessment of, **108**, 155
Quality failures, 13, 52
Quality inspections, for maintenance, 141-142, **142**
Quality issues, 2, 33, 50, 51, 52; decision-making authority for, 16; detectable, 135
Quality management, xii, 2, 7, 8, 9, 13, **19**; brewery practices and, 4; components of, 5; defined, 3, 160; evolution of, 3; focus on, 47; historical context of, **5**; oversight of, 13; practicing, xi, xvii, xviii, 3, 21-26; quality governance and, 15, 16; resources for, 162; responsibilities of, 21-22; role of, 2, 14, 20, 22; as science, 3; skill set for, 9, 32; stagnation of, 25; study of, xvii, 3; system, xviii, 9; understanding, 30
Quality manager certification, resources for, 162
Quality managers (QM), 16, 24, 105; BOK of, 25; decision making by, 20; role of, 10, **10**, 20, 112-113; training for, 50
Quality manuals, 27, 33, 56, 155; defined, 160; policy/specifications/goals and, 17-19; roles/responsibilities and, 19-21; small brewery/example of, 111-119; updating, 111; writing, 18, 21, 26
Quality policy, 16, 112-113; articulating, 18, 19; defined, 160; implementing, 17, 21-26
Quality Priority Pyramid, xv, **28**
Quality program, xvi, 1-2, 26, 45, 47, 48, 59, 64; components of, 27; credibility of, 69; developing, 25; variation in, 33
Quality services, **49**
Quality Subcommittee (BA), 5
Quality system, xii, 1, 2, 15, 16, 22, 103, 104, 122; components of, **28**; defined, 160; eroding, xviii; establishing, 3, 8-9; maintenance of, 21; managing, 18, 153; people/processes for, 154-155; problems for, 10; resources for, 162
Quality System Audit, 103, 104, 106, 109, 153, **154**; parts of, **106-108**

Raw materials, 9, 35, 36, 98, 99; microbiological specifications for, 63; selecting, 5
Raw materials and cardboard packaging sampling plan, **126**
Reactive maintenance, defined, 160
Real extract (RE), 83, 84
Recall coordinators, 117
Recalls: Class I, 98; logs, 116; procedures for, 101, 155; risk of, 13, 14
Record keeping, 47, 57-58, 99, 100
Refractometers, using, 84
Reinheitsgebot (1516), 6

Rejection limits, setting, 45
Repeatability, defined, 160
Reproducibility, defined, 160
Residual extract, 83
Resources, xvi, 48, 161-163; allocating, xiii; managing, 61. *See also* Human resources
Responsibilities, xv, xvii, 15, 16, 19-21, 51, 67, 154; brewery, **10**; GMP, 55; hold/release, 50; matrix of, 11; operational, 9, 21; quality and, 12-14; quality management, 21-22
Return on investment (ROI), 54
Risk assessment, 27, 39, 41, 44, 61, **114**, 156; conducting, 33-37; food safety and, 97-101; formal, 45; HACCP and, 34, 131-133; performing, 62; tools for, 34, 36
Risk priority number (RPN), 36, 135
Risks, 61; analyzing, 2, 34, 62; defined, 160; low level of, 98; prioritizing, 37
Root cause, 25, 31, 57, 67; defined, 160; zwickel sanitation and, 79
Rotenticides, 145
RPN. *See* Risk priority number

Safe Quality Foods (SQF), 50, 104, 161
Safety, 48, 51, 62; gear, 140; impact on, 52; information, 140; issues, 50; product, 35; rules, 24
Sales team, 47, 146
Sampling, 33, 42, 43-44, 78, 79, 88, 159; frequency of, 43; plans, 43, 91
Sanitation, xii, 31, 32, 47, 54-57, 73, 78, 87, 98, 144-146, 156-157; documenting, 56, 100; equipment, 100; facility, 100; planning, 56-57; policy, 113, 114; poor, 79; procedures, 56, 114; understanding, 30; validation and, 57
SCABA, 85
Schlitz Beer, 13, 14
Seals, 51, 115
Sediments, 7, 71, 72, 105
Sensory analysis, 43, 86-90; test kits, 89, 162
Sewage, 145
SG. *See* Specific gravity
Shelf life, xii, 1, 34
Shipping, 47, 51, 105, 154
Siebel Institute, 81
Sierra Nevada, xi, xii

Signage, 100, 145-146
Six Sigma, xvii, 25, 32, 39, 40-41; Black Belt, 31, 50; courses on, 31; defined, 4, 160; focus on, 3; Green Belt, 31, 50; problem solving with, 66; resources for, 162; training with, 39
Skills, 2, 69; assessing, 32; developing, 48; foundation, 30, 31; leadership, 50; management, 30-32, 47, 48; missing, 32; people, 31, 32; problem-solving, 31; process, 48; quality, 9, 28, 29, 32, 49; specialized, 83
SKU, 60, 122
Slide culture test, **75**
Sodium hydroxide, 72
"Some Thoughts on Quality Control" (Brenner), 3
SOPs. *See* Standard Operating Procedures
South African Brewers (SAB), World Class Manufacturing and, 13
SPC. *See* Statistical process control
Specific gravity (SG), 83, 84
Specific process control (SPC), 31
Specifications, 3, 15, 21, 30, 41, 150; charts, 37; codifying, 2; defined, 160; finished product, 116; quality, 7, 155; setting, 17-19, 37-39
Spectral analysis, 85-87
Spectrophotometry, 70, 82, 85-86; AA, 83
Spoilage, 51, 72-73
SQF. *See* Safe Quality Foods
Staff, 11, 28-30, 52; empowered, 19, 24; morale of, 105; quality, 15, 28, 29, 30, 32; training for, 9, 99, 154
Standard, defined, 160
Standard Operating Procedures (SOPs), 49, 50, 57, 104, 106, 109, 113, 128, 154; cleaning operation, 139; development of, 156-157; example of, 139-140; sanitation, 105; writing, 155
Standard Temperature and Pressure, carbon dioxide at, 92
Standardization, 5, 14, 48, 50
Standards, 111; quality, 6, 17, 112
Statistical process control (SPC), 3-4, 5, 8-9; defined, 160
Statistics, 2, 9, 30, 31; quality control, 12, 32
Sterilizers, 70, 79
Strategies, 4, 9, 15, 22, 23, 24, 25; overall, 155; quality, 21
Sulfite concentration, 6

Sulfur dioxide, 85
Suppliers, 9, 155
Support team, 155
Sustainability, 18, 19

Tanks: bright beer, 35, 113; cleaning, 57; fermentation, 29, 39; sampling from, 78
Taste panel ballot, **89**
Taste panels, 87-88, 119; calibration of, 62; full, 90
Tax and Trade Bureau (TTB), 98, 100, 101, 104
TC. *See* Thermal conductivity
Technical brewing, resources for, 162-163
Technical Committee (BA), 5
Technical Quarterly (MBAA), 3
Technology, 11, 69, 92, 95; electrochemical, 94
Temperature, 92; controls, 6, 42, 45, 129; devices, 94; incubation, 79; liquid, 91; measuring, 146; water, 145
Tests, 24, 29, 53, 95; accuracy of, 70, 86; analytical, 43, 87; chemistry, 32, 34, 35, 43, 82-86; conducting, 32-44; discrimination, 88, **88**; duo-trio, 88, **88**; equipment, 32, 53, 70, 91, 156; in-line, 43; innovations in, 75; microbiological, 29, 32, 34, 35, 70-80, 81, 82, 155; non-staining KOH, 72; non-value-added, 69; packaging, 90-92, 94; performing, 69; pH, 80; physical, 34, 35; proficiency, 44; QA, 76, 77, 80; QC, 44, 69, 70, 155; quality, 33, 50, 62; results of, 44-46; sensory, 43, 88, 89; string, 72; student's T-, 9; triangle, 88, **88**; types of/brewery size, **34**; validation of, 101; viability/vitality, 75, **75**
Thermal conductivity (TC), 92, 94
Thermometers, 84, 92
Titration, 70
Tolerance, 37, 38
Total package oxygen (TPO), 91, 94
Total Quality Management (TQM), 3, 12; defined, 160; improvement and, 13; introduction of, 4
Toyota, 8
TPO. *See* Total package oxygen
TQM. *See* Total Quality Management
Traceability, 97, 99, 101, 155; records, 116, 117
Training, 16, 21, 26, 55, 56, 62, 64, 88; continuous, 25; extra, 49; foundation, 30; hands-on, 48; operator-to-operator, 49; plans, 23, 50; problem-solving, 31, 65-66; process, 24, 48, 50, 57; records, 57; resource, 48; scheduling, 52; specialized, 83; staff, 9, 99, 154; statistical, 31; of trainers, 22-23
Training manager, 48, 50
Training manuals, 82
Troubleshooting, 10, 12, 13, 24, 59
TTB. *See* Tax and Trade Bureau
Turbidity, 7

United States Code of Federal Regulations (US CFR), 55, 98
US Bureau of Census, 3
Utensils, cleaning/sanitizing, 55, 144-145, 146
Utilities, 105, 154

Validation, 56, 89; sanitation and, 57; third-party, 101
Values, quality, 14, 18, 19, 28
Valves, 78; manual, 42; sampling, **42**; sanitary, 88
Variation reduction, 3, 7, 34, 39; control chart with, **40**; controlling, 40
Viability, 29, 74, 75, **75**
Visitors, 146, 147; managing, 55
Visitors and Contractors GMP Guidance, 55, 147

Waste, 145; handling, 144, 146
Water, 70; carrying, 145; dilution, 87; pressure, 104; purification, 70
Wet hopping, 60
World Class Manufacturing, 13
Wort, SG of, 83
Wort chiller, 39

Yeast, xviii, 33, 35, 51, 62, 79, 88; cell count, 77; growth of, 81; handling, 6; issues with, 87; quality/quantity of, 29; samples, 74, 75; slurry dilution, 82; staining, 74; types of, 63; viability/vitality, 75; viewing, 71, 72; wild, 105
Yeast pitch, 29, 76, 87

Zahm-Nagel unit, 91, 94
Zeiss, 71
Zwickel technique, 78, 79